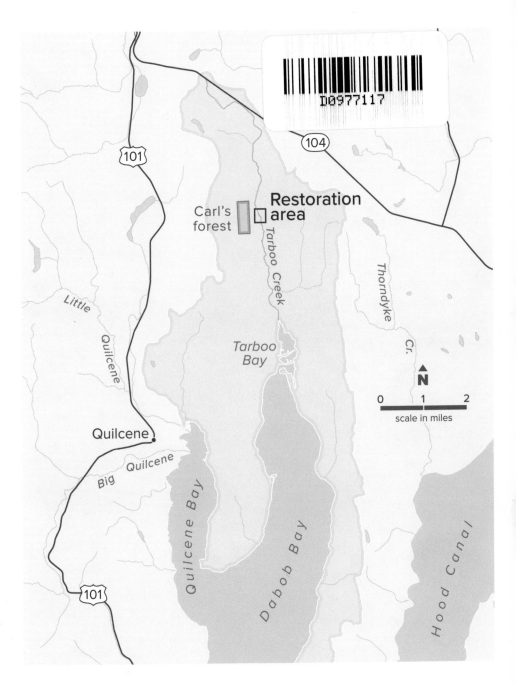

Carl's forest

Restoration area

Tarboo Creek

Thorndyke Cr.

Tarboo Bay

Little

Quilcene

Quilcene

Big Quilcene

Quilcene Bay

Dabob Bay

Hood Canal

0   1   2
scale in miles

N

101

104

TARBOO CREEK WATERSHED

Saving Tarboo Creek

SAVING TARBOO CREEK

One Family's Quest
to Heal the Land

SCOTT FREEMAN

Illustrations by SUSAN LEOPOLD FREEMAN

TIMBER PRESS · PORTLAND, OREGON

Published in 2018 by Timber Press, Inc.
The Haseltine Building
133 S.W. Second Avenue, Suite 450
Portland, Oregon 97204-3527
timberpress.com

Mention of trademarked products does not constitute a guarantee
or warrantee of the product by the publisher or author and does not
imply its approval to the exclusion of other products or vendors.

Printed in the United States
Text design by Adrianna Sutton
Jacket design by Mumtaz Mustafa and Adrianna Sutton

Library of Congress Cataloging-in-Publication Data

Names: Freeman, Scott, 1955– author.
Title: Saving Tarboo Creek: one family's quest to heal the land /
by Scott Freeman; illustrations by Susan Leopold Freeman.

Description: Portland, Oregon: Timber Press,
2018. | Includes bibliographical references and index.

Identifiers: LCCN 2016055648 | ISBN 9781604697940 (hardcover)

Subjects: LCSH: Nature conservation—Washington (State)—Olympic
Peninsula. | Human ecology—Washington (State)—Olympic Peninsula. | Tarboo
Creek (Wash.) | Leopold, Aldo, 1886-1948. Sand County almanac.

Classification: LCC QH76.5.W2 F74 2018 | DDC 333.7209797/94—dc23 LC
record available at https://lccn.loc.gov/2016055648.

Men nowhere, east or west, live yet
a *natural* life, round which the vine clings,
and which the elm willingly shadows.

—HENRY DAVID THOREAU
A Week on the Concord and Merrimack Rivers
(emphasis original)

Contents

~~~~~~~~~

Dedication

My father-in-law, Carl Leopold, was born on December 19, 1919. Less than two weeks before his twenty-second birthday, the Imperial Japanese Navy attacked Pearl Harbor. His life, and the lives of millions of other American men and women in their late teens and twenties, changed that day. The United States declared war on Japan on December 8, 1941, and on the Axis powers led by Germany on December 11.

On Christmas Day 1941, the free world essentially consisted of Australia, Canada, Great Britain, New Zealand, and the United States. The government of every other country was a dictatorship or fascist, with the exception of Sweden and Switzerland, which remained neutral in the war against totalitarianism. The Battle of Britain had been won, but everywhere else—in Eastern Europe and Russia, Africa, East Asia, Southeast Asia, and the Pacific —the armies of tyranny were advancing.

Imagine that you were a high school or college student on that morning. If you were living in Europe or Asia, you were under occupation. If you resisted that occupation or happened to be Jewish, Romani, a Freemason, or homosexual, you were almost certain to be killed.

What if you were living in what was left of the free world? My father-in-law and his brothers and sisters celebrated Christmas that year in the living room of their home in Madison, Wisconsin. Life there was quiet, but there was also disquiet: the rest of the world was in flames. The young were realizing that they had to fight a war—one that the democracies were losing badly—before they had any hope of finishing school, starting a family, and getting a career under way. They were being called upon to save the world.

My father-in-law enlisted in the U.S. Marines and spent four years in the Pacific as a captain in the artillery. He survived, went back to school in 1945, and had a long, distinguished career as a research scientist. But two-thirds of the young men in his officer training class were killed in action. This book is dedicated to them, and to him.

Today, seventy-five years after the Second World War became a global conflagration, high school and college students are again being called on to save the world. This book is dedicated to them, too.

Introduction

~~~~~~~~~~~~~~~

Noticing Things

My uncle Carl Holtz farmed in southeast Wisconsin for forty years. But before he started farming, he went to the University of Wisconsin to row on the crew team. While he was a student there he took a course on wildlife biology—then called game management—from a professor named Aldo Leopold.

During the semester, each student was required to have a brief one-on-one conversation about the course with Leopold in his office. More than twenty-five years later, my uncle told me about that meeting: "I sat there like the dumb jock I was back then, you know. Professor Leopold was asking me about this and that, and I had absolutely no idea what he was driving at. But then something caught his eye out the window, behind his desk. He looked at it for a moment, then turned to me and asked, 'Carl, what bird is that?'"

"I had no clue, of course," he laughed. "But years later I realized it was a palm warbler, migrating through."

My uncle was a big man, with hands the size of salad plates. He held them up. "Leopold knew I wasn't going to go on to graduate school or become a wildlife biologist. He just wanted me to look up and notice things." Uncle Carl put his hands down and nodded at me. "And so I have—ever since."

~

Outside my window in Seattle right now, a flock of bushtits is feeding in the bare branches of a birch tree. Some are upside down; some are right-side up. They are flitting, fluttering, jumping. Then they disappear all at once—diving into the cover of a nearby Douglas-fir tree. Now they're back. A moment later, they're gone—until tomorrow.

These birds are adults and juveniles. They are neighbors from the previous year and new immigrants to the neighborhood, and by now are well acquainted. The members of a winter flock like this one find each other in late summer and stay together until the following breeding season. Although bushtits dominate this particular group, there are also some golden-crowned kinglets and at least one chestnut-backed chickadee. Around here, it's common to find northern juncos, black-capped chickadees, and hairy woodpeckers in the mix, and sometimes even ruby-crowned kinglets.

You can find these types of mixed flocks almost anywhere you go in the world. In Japan, there would be marsh tits and great tits and goldcrests—close relatives and look-alikes of North America's chickadees and kinglets. The Eurasian treecreeper and Eurasian nuthatch would take the place of our brown creeper and red-breasted nuthatch; Japanese pygmy and great-spotted woodpeckers would stand in for our downy and hairy woodpeckers.

In the lowland rainforests of southern Ecuador, all Hades breaks loose. There may be twenty-five species and forty or more individuals in a mixed foraging flock like this. In addition to woodpeckers and woodcreepers, there will be several types of antwrens, a handful of flycatcher species, and a bouquet of tanagers: yellow-throated, blue-winged, orange-eared, blue-browed, and bay-headed, among others. The colors streak from branch to branch. They are dazzling, brilliant, sublime.

Typically, each species in a mixed flock will be eating something slightly different, in a different part of the vegetation. Out my window, the bushtits glean from the tiniest twigs; chickadees pick at branches; brown creepers probe the trunk's furrowed bark; hairy woodpeckers rap at spongy, rotting spots in the wood.

When these little gangs appear, moving slowly but steadily through the trees, the woods look like Central Park on a summer Sunday. You'll find every size, shape, color, and linguistic group imaginable among birds—all moving and jostling, going about their day.

For a mixed flock like this, there is knowledge in numbers. Large flocks can draw on the collective wisdom of fifteen or twenty memories, finding food in obscure locations when ice and snow coat the branches and ground.

There is safety in numbers, too. If a sharp-shinned hawk dove into this birch tree and surprised the group, the little birds would scatter like shot—making it hard for the predator to draw a bead and snatch one from the air. And to avoid surprise, many eyes are better than two. In black-capped chickadee flocks, individuals that notice flying predators give a high-pitched "seet" call; in response, the others dive for cover. But if the predator is sitting, the spotter gives the "chick-a-dee" call and adds "dee's" to indicate the degree of danger. Biologist Chris Templeton and co-workers figured this out by bringing live predators into a large outdoor aviary where a chickadee flock was living. Chickadees are little—almost as tiny as bushtits—and it is the small, agile killers like saw-whet owls and northern pygmy owls that worry them the most. In the experiments, small predators could elicit a string of five "dee's" or more. But big, lunking hunters like great gray owls, which strike fear in the hearts of snowshoe rabbits and grouse, got only a "dee" or two—barely more than the response to a harmless, seed-eating bobwhite quail. Follow-up work by other biologists showed that Carolina chickadees do the same thing.

Later, Templeton and Erick Greene showed that red-breasted nuthatches, which are common in chickadee flocks, respond much more strongly when they hear long strings of "dee's" as opposed to short strings. The nuthatches understand Chickadeean, even if they can't speak it. After this work was published, researchers found the same pattern in mixed flocks from locations around the world. In most or perhaps all cases, alarm calls are mutually intelligible to mixed-flock members.

So the next time you're out and about and hear chickadees giving the call that inspired their name, count the "dee's." The birds may be describing how dangerous you are. Or if early in the morning you hear males giving territorial "fee-bee's" back and forth, appreciate that each individual's voice has unique and recognizable charac- teristics—even though it's a simple, two-note song. To us, they all sound alike. But chickadees know their neighbors by voice.

As you walk, then, you'll be doing the same thing that bushtits are reminding me of today and that Aldo Leopold prompted my uncle to do sixty-five years ago: look up and out. When you begin to notice the wild things around you, like the clash of orange and yellow on the head of a golden-crowned kinglet, even a dull day can spring to life.

⁓

This book is about small things, like watching birds brighten a Feb- ruary afternoon or hearing a tree frog on a cold, moonless night; about planting a tree or hoeing beans. But it's also about a big thing: what life will be like in 2100, when the human population passes 11 billion and almost half of the species alive today are extinct; when every organism left on the planet will be trying to cope with what may turn out to be the most rapid period of climate change in Earth's 4.6-billion-year history.

~

The twentieth century was rife with conflict among nations and races. Tens of millions died, and hundreds of millions more suffered, in three world wars—two hot and one cold. Those deaths were tragic, but they also accomplished something profound: fascism, communism, and colonialism are gone, or nearly so. Racism, nationalism, tribalism, and religious extremism are still alive, but they are not well, relative to historical norms. The -isms that have plagued our kind for thousands of years no longer move hundreds of millions to violence and exploitation.

As the twentieth century's wars were being fought and won, the flowering of modern science and engineering created wealth and improved health. In a hundred years, we established standards of living that had been unthinkable since the dawn of humankind some 150,000 years ago. Mothers could remain calm if their children came home with a slight limp, because it wasn't polio. Tens of millions of us ate like kings, wore fine clothing, and lived in small castles.

The challenge of this century is a product of the past century's successes.

If you are young as you read this, you are entering a world filled with danger: peril from those who cling to the old ways of tribalism and religious extremism, and new threats to our planet's ability to support life, brought about by our species' recent success at acquiring and using resources. If you are a parent or a grandparent as you read this, your children and grandchildren are entering a world you did not experience but must prepare them for.

As far as we know, no human culture has ever limited its use of resources voluntarily. Throughout history, people have used forests, wildlife, water, and soil until they were used up. When that day arrived, the individuals who were affected either died or moved.

Time and again, overexploitation led to the collapse and disappearance of entire civilizations.

The power of those lost civilizations to devastate landscapes was nothing compared to ours, however. We have the technological prowess, the numbers, and the appetite to threaten the future of most life forms, including our own. And now there is nowhere to move to. Earth is full.

So yes, there is danger. But there is also possibility.

In 1949 my uncle's teacher, the forester and wildlife biologist Aldo Leopold, published a book of essays called *A Sand County Almanac*. It comprised nature observations, hunting and fishing stories, and an indictment of the gospel, unquestioned at the time, that destroying natural areas in the name of human progress was good. But most important, it also contained an idea—a solution to the waste and loss that occurs when we treat the land around us as property, as something we are free to use and abuse without restraint or consequence. Leopold proposed a fundamentally new way of thinking about the relationship between people and the natural world. His idea was simple: good people should treat the land around them the same way good people treat the people around them.

To Leopold, this insight was a logical step in the progressive evolution of human ethics—a process that started thousands of years ago, when the first inklings of civilized behavior began to rein in our predisposition to act out of pure, unbridled self-interest. At the genesis of this ethical sequence, the novel idea was that we should act with care and consideration toward members of not just our immediate family but also our immediate community—our neighbors. Leopold saw the Mosaic Decalogue—the Ten Commandments—as a summary statement of this early stage, a marker and touchstone that we still use today. Later, the Golden Rule and the Sermon on the Mount extended the standards of ethical behavior to members of our broader society—to people we do not know and to those outside of our tribe or group. The invention and practice of democracy applied these norms to the community called a nation. Although Leopold died in 1948, I would claim that the international organizations and trade agreements that have emerged since the Second World War— the Association of Southeast Asian Nations, the European Union, NATO, the Organization of African States, the United Nations—are attempts to extend ethical standards to the community of nations. Since humans first began to walk upright and live in groups larger than extended families, our circle of ethical responsibility has been expanding outward.

Leopold simply extended this progression to the natural world. In his view, ethical people would no more harm a marsh or forest than they would harm their family, or their country, or any human being who wasn't actively trying to injure them. Leopold called this idea the land ethic. It was an eleventh commandment: thou shalt not harm the land.

The key idea was that land cannot be your property, any more than another person can be your property. People do not have the right to use land and resources badly any more than they have the right to use other people badly. Leopold wrote *A Sand County Almanac* because he saw no other way to stop the thoughtlessness and destruction that was occurring everywhere he looked. The only lasting solution had to come from ethics—from the depths of the human heart.

But in the same breath that Leopold proposed this simple idea, he also wrote that "nothing so important as an ethic is ever written . . . it evolves in the minds of a thinking community."

Saving Tarboo Creek is written for that thinking community—for the network of people across the globe, from all walks of life and every possible background, who live a thoughtful, examined life. An examined life pursues wisdom. It is mindful of the needs of others. It is a life worth living.

My wife, Susan, is Aldo Leopold's granddaughter, and our two boys are his great-grandsons—part of the fourth generation of Leopolds to work on ecological restoration. The chapters that you'll be reading tell two stories about this legacy, woven as on a loom. The warp—the long, vertical threads that a weaver sets to begin a rug or tapestry—is the work that we are doing as a family to help restore a heavily degraded salmon stream that runs into the Pacific Ocean off the northwest coast of North America. The weft that shuttles back and forth is made up of strands from the larger world that affect the work: the history of resource use over the past two thousand years and observations on the current and projected state of the world. The warp is what we do; the weft explains why. As we plant trees, build trails, and watch our little creek change over time, we consider the tasks that we are doing together and the world around us. This is how we take part in the thinking community that Susan's grandfather wrote to.

The state of the world defines a twin challenge for us all. The first is to discover a set of values based on self-restraint and on a commitment to the long-term health of human and natural communities. The second is to live by them. The goal is to find a harmony with each other and with the land.

It has never been done before. But that doesn't mean it can't be.

A Stream Is Born

In the Puget Sound Basin of western Washington, stream restoration work starts in early August. The key is the rain—it's rare from early July through mid-September, and soils dry to powder. Rivulets dry up; creek depths drop to inches; rivers slow to a warm, sluggish crawl.

The salmon that call Puget Sound home leave their foraging grounds in the northeast Pacific Ocean about this time and home in on their natal streams. Fish with return addresses in big rivers may run up into the freshwater and spawn right away, but fish from small rivers and creeks usually delay. In the low waters of September, the small-stream fish would scrape their bellies over gravel for mile after mile; they'd have no bellies left by the time they reached their spawning grounds high in the watersheds. They'd also have no place to hide from the bald eagles, black bears, and river otters that are checking the streams every day, anticipating their return. So they wait.

Then, November arrives. Temperatures drop to the low 50s, winds rise, the first storms blow in from the north Pacific, and a salmon's thoughts turn to love. The fish will still hold back, schooling in bays and estuaries, until a howling storm or two have swollen the smaller rivers and creeks. As the water deepens, the salmon make a run for it—moving pell-mell upstream. They swim against the current, in the manner of poets and prophets.

～

The little salmon stream that my family is helping to restore is located on Washington's Olympic Peninsula and is called Tarboo Creek. The name is pronounced like two beats on a Suquamish drum, with slightly more emphasis on the first—a gentle TAR-boo. When people say tar-BOO, they scare the fish.

Tarboo Creek's waters flow from north to south and empty into Tarboo Bay—a tiny fiefdom in the Pacific Ocean's vast empire. Tarboo Bay runs into two long sand spits, then broadens into Dabob Bay, which opens into Hood Canal, a fjord that forms the southwest corner of Puget Sound.

To the south, Hood Canal ends near the steps leading to the state capital building in Olympia, Washington. The canal and the rest of the sound's western shore are sparsely settled, but its east coast is home to more than 2.5 million people. There, bays that are large enough to support ports attracted the European settlers who founded the cities of Bellingham, Everett, Seattle, and Tacoma.

To the north and west, the sound connects to the Strait of Juan de Fuca—a conduit to the open Pacific that is an international boundary, separating Washington's Olympic Peninsula and British Columbia's Vancouver Island. If you put a kayak in the Strait of Juan de Fuca and paddled straight west, you would eventually run aground on Sakhalin Island, off Siberia's southeast coast.

Tarboo Bay is one of the highest quality estuaries left in Puget Sound. Unlike most of the region's coast, the bay's shoreline has a minimum of homes, breakwaters, docks, and jetties; the water quality is good enough to support natural spawning by Pacific oysters. The bay drains at low tide, creating an expanse of mudflat dotted with glaucous-winged gulls and Northwest crows. Tarboo Creek flows out toward the spits in a winding channel through the sand and muck, chasing the retreating salt water. But at high tide, the sea fills the bay to brimming. Ducks bob; salmon swim; seals hunt.

In November, when the salmon follow a high tide into a creek channel deepened by the first rains, they enter one of the last tiny patches of low-elevation old-growth forest left in Puget Sound. I've seen aquatic biologists fall into a reverent silence there, marveling at the cool, even flows of clear water rippling over gravel beds and wending around sandbars, before stooping like eight-year-olds to turn over rocks and begin cataloging the scores of midge, caddis fly, and mayfly larvae that cling to the cobbles. The gradient is low and the channel wide, lessening the chances that storm-driven floods will scour the bottom and rip out salmon nests. There are jumbles of mossy downed logs and jangles of root wads, creating fish-filled pools and eddies. Bracket fungi peek out from the wet snags; they are releasing nutrients into the creek, at a slow drip, as they rot the downed trunks.

Sword ferns cover the sandy deposits that line each bank. Above them, sprawling vine maples form a shrub layer; higher up, a smattering of small red alders holds sway. Over everything, craggy old bigleaf maple trees and towering Sitka spruce form a canopy. The maples are covered with moss and dripping with lichens; the spruce are massive—6 feet or more across at eye level, with trunks that rise five stories high at the first limb and then disappear in a tangle of needles and branches, perhaps topping out at 175 feet or more. The biggest Sitka of all has a root 3 feet in diameter that buttresses away from the base and bridges a tributary to the creek. You can sit on that root on a wet November day and watch chum salmon swim under your feet—headed up the tributary to spawn.

We can only guess, but the tree may be more than 750 years old. It is a great-great grandmother, many times over.

Walking the banks of the creek, you can find cougar, bobcat, or bear tracks, or at least the spindly impressions from a great blue heron. As you walk away from the stream into the uplands above the floodplain, you begin wending your way around pillars of reddish-brown western redcedar trunks. There a lacy canopy of drooping boughs casts enough shade to limit the undergrowth to a few scattered ferns.

At first the woods look like they could go on and on, but they are actually just a 160-acre preserve. To the north, things get difficult from a salmon's point of view. For much of its 7.5-mile length, Tarboo Creek was ditched or diverted and the surrounding wetlands drained. A series of badly installed culverts blocked fish passage to large portions of the main channel and tributaries. If the culverts that carry a stream under a road or driveway are too small or placed at the wrong grade, two things happen. After storms, large flows give them the look and feel of a fire hose, with so much water traveling so fast that salmon can't fight their way up. At the downstream end, the torrent scours the surface below the lip of the pipe, eventually creating a waterfall and perching the culvert 3 feet or more above the rest of the stream. This is too high for fish to jump, even the athletic coho salmon. A poorly designed culvert can block migrating fish as effectively as a net.

⁓

An old gentleman who grew up in a house near the creek said that as a boy he used to lie awake in his second floor bedroom, unable to sleep for the splashing of salmon spawning in the distance below. By the late 1990s some chum and coho were still nesting in Tarboo Creek, along with a smattering of sea-run cutthroat trout, but their numbers were a fraction of historical norms. And steelhead—rainbow trout that breed in freshwater but spend most of their life in the ocean—had been extirpated decades earlier. More than seventy years had passed since the stream had seen salmon runs like the old man described.

In 2004 our family bought an 18-acre parcel that straddles the main channel of Tarboo Creek. When we did, we joined a community of individuals and organizations working to reforest abandoned pastures and degraded wetlands in the valley in hopes of restoring Tarboo Creek as a high-functioning salmon stream. Most of the individuals involved are private landowners in the watershed; the organizations include county, state, and federal government agencies, local Indian tribes, and private nonprofits like the Jefferson Land Trust. The entire project is coordinated by salmon biologist Peter Bahls, who directs the Northwest Watershed Institute (NWI).

By the time we got involved, the project was already gaining momentum. The 160-acre tract that includes that patch of old growth near Tarboo Bay had recently been purchased by the State of Washington's Department of Fish and Wildlife. The acquisition went through just before the tract was scheduled for clear-cutting by a private forest products company. The biggest Sitka spruces would've been spared—forest use regulations require loggers to maintain a 50-foot setback near fish-bearing streams—but the cedars on the bluff would've been felled for utility poles or decking and replaced with a plantation of fast-growing Douglas-fir cultivars. The cut was delayed at the last minute when Peter Bahls discovered an osprey nest and notified the Department of Natural Resources, which oversees logging. It's illegal to cut forest near an active nest of a protected bird-of-prey species; the osprey discovery bought enough time for the Department of Fish and Wildlife to budget the money for the purchase.

By 2004, Jim Yeakel and Joan Purdy, who run a farm and a cabinet-making business not far from the mouth of the creek, had worked with NWI to regrade and replant a rivulet that connected the main stem of the creek to a pond on their land—ideal habitat for juvenile salmon to grow in before returning to the ocean. Near the headwaters, Alan Iglitzin—a violist who directs chamber music concerts each summer at the Olympic Music Festival—had helped NWI remove a culvert that was blocking salmon passage and replace it with a wooden bridge. In between, the county highway department had replaced two other old culverts that were preventing salmon from moving upstream.

So when we arrived, the mouth of the creek was protected and the worst of the blockades had been removed. The creek was reopened for business. Salmon could move into habitat that had been closed for more than fifty years.

Our land is about halfway between the headwaters and the mouth of the stream, and was the next project on the list. The watercourse there looked like an open wound—it was easily the most badly degraded stretch of stream in the watershed. The creek had been channelized through our property in the early 1970s. Thirty-five years later, the water was shooting through a steep-sided, arrow-straight ditch, excavating the bottom as it went. In the worst sections we could stand on the gravel streambed and not see out. The creek was 6 feet under. It was a sluiceway conveying sediment that would slowly choke the life out of Tarboo Bay.

In what was left of the creek's floodplain at our place, thistles and horsetail were chin high. Himalayan and Eurasian blackberry formed an archway over most of the creek channel, interwoven with tangled mats of reed canary grass. All three plants are nonnative and invasive, and are listed as noxious weeds in western Washington. The blackberry canes, in particular, are armed and dangerous. They brandish centimeter-thick stems studded with spikelike thorns and can grow 10 yards or more a year.

The previous landowners had also cut all of the 3-to-4-foot-diameter western redcedars that were growing on the hillside above the ditched creek. The cut left a smattering of spindly young cedars—whips in logger parlance—along with a stand of bigleaf maples. Foresters call this approach high-grading, or thinning from above. It's a way to cash out quickly. If you thin from below, you remove smaller trees that are competing for light and water so that the bigger trees can get even larger over time, and more valuable. Thinning from below is a way to manage for longer-term, sustained yields.

Fortunately for us, the owners had left the large bigleaf maples—apparently not realizing that they can contain patches of compression wood, which grows in response to the weight of leaning stems or branches. In maples, compression wood can be figured with dazzling swirls and streaks that are prized by luthiers for making violin and guitar backs. People in the business claim that a large maple with extensive figuring can be worth $5,000 or more on the stump. It's common for maple poachers to patrol back roads on the peninsula, looking for big specimens. If the coast is clear they'll drop the tree, find sections with figured wood, and cut them out, like ivory.

The 18-acre parcel we bought was actually part of an 83-acre farm the previous owners had purchased. After logging it—clear-cutting 40 acres and high-grading the larger redcedar off our place and the big Douglas-fir off the farm next door—they subdivided it into three parcels and put them up for sale, all within a year of their original purchase. In this, they were following a long tradition. When Lewis and Clark returned to St. Louis from the Pacific coast, they turned in their notes describing the natural wonders the Corps of Discovery had seen—an Eden of rolling prairies brimming with wildlife, pristine rivers and lakes, and cathedral-like forests that stretched for hundreds of miles. Then Meriwether Lewis began speculating in land and the fur trade. He wanted to cash in.

At our place, the previous owners had even dug two ponds and bulldozed a house site and circle drive, thinking they would sell to wealthy retirees. Instead the place sat, attracting almost no interest. It had been for sale for four years when we came along.

～

An ecologist I know advises new landowners to wait five years before starting restoration projects. He wants them to walk their land—to watch it through the seasons and over the years. The idea is to learn a place before starting to change things. This is sound advice, but we couldn't take it. The Northwest Watershed Institute had a grant to dig a new, meandering channel for the stream, and all the required permits were in hand. The idea was to make crooked that which was straight.

Stream restoration permits usually start August 1 and expire at the end of September. The interval is timed to coincide with the late-summer drought that is typical of the region, when heavy rains are unlikely to erode newly dug channels. The regulatory agencies also want you to wait until the salmon fry of the year are big enough to handle the disturbance. Then the challenge is to get the work done and the machines out of the water well before the rains begin and the adults return to breed.

In a floodplain, the purpose of a ditch is to collect surface water, drop it down, and get it to run off as fast as possible. A well-engineered, well-maintained ditch will dry the soils enough to make pasturing or plowing possible. But in the Tarboo watershed, the channelization and pasturing and plowing weren't successful. The last dairy farms in the valley closed down in the 1960s—victims of poor agricultural soils, steep slopes, and competition from large factory farms. The ditching at our place was a last-gasp effort that didn't change the outcome. A century of trying to grow livestock, instead of big trees and big fish, had ended in failure.

The purpose of a stream restoration is to undo everything that channelization does: to bring the water back up to the surface, slow it down, and wet the soils. Once the stream is back at ground level and meandering its way seaward, the groundwater level will rise and soils will resaturate, supporting the growth of wetland plants. A ditch is straight, narrow, and predictable; a stream is curvaceous, broad, and fickle. Other than deepening, a straightened channel may not change appreciably in half a century; a free-flowing river changes almost constantly.

To introduce complexity in a restored salmon stream, you have to create quiet, backwatered pools where the current is slow; rapid, noisy riffles where adult females like to lay their eggs; and everything in between. The pools are where the salmon fry will congregate, burning up a minimum of energy to maintain their place in the current and grazing in peace on organic debris and invertebrates, with hidey holes nearby in case a kingfisher or great blue heron visits. Riffles are preferred spawning sites because the current cascades over gravel, creating white water that oxygenates the eggs stashed below.

Restorationists also want to install just as many logs as possible. The best salmon streams in the Pacific Northwest run through dense stands of towering old trees and are pierced and prodded and poked with windfall timber. The moss-draped logs create pools where both young and old fish can hide from predators; the decaying wood drizzles nutrients used by aquatic plants and insects, possibly ending up in young salmon.

Peter Bahls and civil engineer Tom Smayda studied intact stretches of Tarboo Creek to figure out the size of the stream's normal meanders, and they used recent weather and water flow records to model storm-driven surges. The research let them work out how deep and wide to make the channel—what its new profile would look like. They were designing for high water and for flooding, calculating how often the water would overtop the banks and where it would go when it did.

The result was sheaves of blueprints. The ditch through our place ran about 1,000 feet; the remeandered stream would run 1,500. And it would have features: new meanders, logs, root wads, pools, and riffles and mini-rapids lined with imported gravel. It was all laid out, in blue and white.

Susan's uncle Luna Leopold had spent a career studying how streams work and was widely considered to be a world authority on river hydrology. He critiqued the initial drawings on what turned out to be his deathbed. When he and the state permitting agencies had signed off on Tom's revisions, we had a project.

◠

The remeandering crew was made up of our family—including two sons, a sister, her boyfriend, a niece, Susan's father, and her aunt Estella Leopold—along with Peter Bahls and Sean Gallagher from NWI. But the main man was Bob Harrison.

For nine months of the year, Bob makes his living digging basements for suburban tract homes. But for three months of the year, he remeanders salmon streams. He smokes two packs of cigarettes a day, has a belly like a Buddha, and is the go-to guy for salmon stream restoration in northern Puget Sound. He's a fisherman, so he knows how good streams look and act, and he understands his role: "We're just trying to give the creek good bones," he likes to say. "Then it'll make itself over time."

Bob masters his machine. Most excavator operators can dig; Bob can sculpt. After watching him work his big orange Hitachi for a couple of hours, Susan turned to me and said, "If I could just put a brush in that thing's claw, he could probably paint pictures with it."

For two weeks, Peter Bahls staked out the sinewy new channel and Bob dug and scraped and contoured. Peter surveyed to check depths and widths; Bob adjusted both to match Tom's drawings. Then he brought in long logs—damaged goods scavenged and trucked in from nearby clear-cuts—to place in the new channel. When a particularly big cedar log dropped into place, Bob would snatch the cigarette from his mouth and whoop like a cowboy. "Hot damn!" he'd holler. "The fish'll love *that*."

Once the wood was in place and the excavator moved on, the field hands moved in. Sean led; we followed. As we worked and got to know Sean, we learned that he lived in the valley and had been working with

NWI for three years, helping with restoration projects, doing spawning surveys to count and map salmon nests during the run, and sampling the creek throughout the spring and summer to count salmon fry. We also learned that he is of Irish-Italian-Inupiat (Eskimo) extraction. His maternal grandmother was a native of King Island, Alaska, a now-uninhabited rock in the Bering Straits. In 1778, King Island was home to two hundred people who looked out of their walrus-skin residences to see Captain James Cook sail by. Cook mapped the island and gave it its English name. Much earlier still, King Island had been part of the Bering land bridge. Sean's ancestors may have been living on that rock since people first walked into North America from northeast Asia, more than fifteen thousand years ago.

When Bob was done, the banks of what would become a meander were raw and the bottom of the channel dry. Our first job was to spread seed from rye and other annual grasses above and just below the future water line and add a layer of hay. By covering the bare soil, the hay would lessen the impact of rain and prevent little rivulets from forming and carrying silt into the channel. It would also mulch the grass seed, keeping newly sprouted seedlings from drying out and dying later that summer.

Then we rolled out and pegged down a net of heavy jute twine called coir (COY-er). This was a key step: the coir would prevent the new stream from cutting into the raw banks, preventing erosion until the new vegetation could root and hold the soil. In time, the coir would decay and disappear.

To keep the fresh, stiff coir in place, we staked it with metal landscape staples and lengths cut from nearby Pacific and Scouler's willows. Willow stakes sprout if conditions are right and would eventually produce thickets that help shade the creek. An arch of leaves and branches is good: it keeps the water cool, shields little salmon from kingfishers and herons, and drops organic material that supplies nutrients. We were also hoping that the willows would provide enough food to attract beavers.

When we had a stretch of creek bed prepared like this, Bob would come rumbling back in the Hitachi and use the excavator's bucket like a salt shaker, sprinkling gravel to a depth of 3 to 4 inches all over the channel. Female salmon use their tails to dig nests—called redds—in gravel beds. When a female releases eggs into the hollow she's excavated, the attending male releases sperm. Timing is everything; fertilization occurs as the eggs settle into the nest.

The females choose nesting sites where the gravel is small enough to move with a thrashing tail but large enough to cover the eggs efficiently, and where the water is moving fast enough to keep the developing embryos well oxygenated but not fast enough to carry the eggs downstream before they settle into the nest, or threaten to scour the redd after the eggs are covered with gravel. It's not surprising that the females are good at what they do; they've been at it for a while. The first salmon appear in the fossil record about 20 million years ago.

We field hands kicked and shoveled the gravel to cover the bottom of the new meanders, gave the coir a few tugs to make sure it was held tightly, double-checked to make sure all the bare dirt had a layer of seed and hay, and moved on. The restored stream channel was ready for water.

~

After a week of work, it was time to divert the creek from the ditch and let it start winding its way through the new meanders. The remodeled streambed was finished and connected to the downstream end of the ditch, and now the task was to block the ditch at the upstream end so the water would divert and run into its new channel.

To do this, we stacked a pile of sandbags next to the upstream point where the ditch and the freshly dug watercourse met. Then three of us filed into the ditch and stood shoulder to shoulder, facing upstream. Working quickly, we laid black plastic sheeting in the stream. As the water began to pool against the sheet, another group bucket-brigaded sandbags to Peter Bahls, who threw them down in a line to hold the plastic flush across the bottom of the ditch. We tugged at the top of the plastic sheet, struggling to hold it fast against the pressure of the damming water. The rest of the crew hustled bag after bag down, building a wall that gradually began taking some of the pressure off the black plastic and us.

As the sandbag dike got taller, the water rose behind it until it reached the level of the new channel. That was the tipping point. First one tiny rivulet, then another poked out into the meander and soaked into the gravel. Others followed, eventually converging into a single column that began snaking its way out of the ditch and into the restored channel, wetting more and more gravel and coir. Susan and her sister took pictures, as if at a christening.

As we finished the sandbag wall, the water level rose until a steady stream was flowing through the virgin bed. Two of the cousins ran alongside the water, whooping and laughing and shouting words of encouragement. It was like watching an innocent man walk out of prison, exonerated, after thirty-five years. You follow him, trying to imagine what this freedom must feel like, and cheer him on.

In the meantime, Bob's excavator was throwing dirt into the ditch behind the temporary sandbag-and-plastic dam, the bucket wheeling back and forth frenetically. When Bob finally had a chest-high wall built behind us, the three of us released our grip on the black plastic sheet. We crawled up and out of the ditch massaging our whitened knuckles, trying to work the cramps out of our hands.

Bob kept filling the ditch behind the sandbag dam. When it was solid enough to stop the flow on its own, we took the sandbags back out, one by one, and retrieved the plastic sheet. Just downstream of where Bob was working, the ditch was beginning to dry up. The creek had a new home.

A cluster of us set to walking up and down the ditch, finding freshwater mussels that were stranded in the drying muck. When we spotted one, we dug it out with our fingers and put it in a bucket with water. Freshwater mussels burrow like saltwater clams. They probe the substrate with a muscular foot and pull themselves down, until a crease at the other end of the shell is just at the surface of the stream bottom. That crease opens into a siphon; thousands of cilia on their gills, deep inside the shell, beat in unison to create a current that flows through the siphon's intake tube, over the gills, and out the siphon's exit tube. The gill filaments, layered in overlapping plates, trap bacteria and other microscopic food particles as the current passes through, while oxygen diffuses from the stream water into the bloodlike hemolymph. So we looked for the little black mouths filtering water as we walked up and down the ditch, and found individuals ranging from a half dollar to a pocket knife in length. When a bucket was full, we transplanted the mussels to the new channel—laying them on the gravel so they could dig their way in.

Although North America has the highest diversity of freshwater mussels of any region in the world, mussels do not do well in polluted water or silted-in streambeds. The Nature Conservancy estimates that 70 percent of native mussel species are extinct or endangered in the United States. If a creek has strong populations of freshwater mussels, it means that the water quality is good. We started to keep track of the mussels we found to transfer but gave up when we tallied more than a thousand in a 50-yard stretch of drying ditch. They were too abundant to count. If mussels are as reliable an indicator as they are purported to be, water quality in Tarboo Creek is exceptional.

As the mussel rescue continued, Peter Bahls and Susan waded up and down the old ditch in hip boots, netting salmon fry that were stranded so we could ferry them over to the new channel, along with the mussels. Sean and various friends and relations collected equipment.

Estella Leopold had returned to work several days before, but Susan's father, Carl, was there to the end, helping us transplant mussels and salmon fry. He was eighty-five years old at the time.

~

In 1935, when Carl was fifteen, about our younger son's age, his father bought 80 acres of abandoned farmland in central Wisconsin for back taxes—about $8 an acre. The family spent the next twelve years of weekends and vacations there; Carl was present for seven until the war broke out. They remodeled an abandoned chicken coop into a kitchen and bunkhouse they called the Shack, planted tens of thousands of pine trees, and restored native prairie vegetation. The work the family did on that farm was the first attempt that had ever been made at ecological restoration.

Carl's father, Aldo Leopold, wrote *A Sand County Almanac* in an attempt to explain why the family spent all of their weekends and vacation time doing stoop labor together. Originally he'd talked about buying the farm as a place to hunt grouse and ducks. But that wasn't really what he had in mind. Working on the Shack was his attempt to put the land ethic into action.

The land ethic is an extension of a moral guideline that is the cornerstone of virtually every enduring ethical and religious system. In the Judeo-Christian tradition, it is the sum of the law and the prophets, encapsulated in a core declaration from the Sermon on the Mount: always treat others as you would like them to treat you. *A Sand County Almanac* simply asked people to extend this ethical behavior to the organisms that share the planet with us. It's an idea that people from North America's First Nations have had for a long time.

It is essential to realize, though, that both ethical commitment and religious faith grow through practice. They have to be realized somehow. Ethical behavior can be lived through caring for the sick and needy; religious understanding can grow through prayer or meditation or communion. One way for a land ethic to grow is through restoration. Working at the Shack was the Leopold family's way of treating the land we live on with the same respect we show to other people.

Carl's sister Nina recalls that when her dad first told the family about the land purchase, she had visions of a place in the country with geraniums and a white picket fence. But the farm Aldo bought had been badly abused. The original vegetation had been stripped and the soils bled dry by fifty years of poor farming practices. The previous owner had burned the house down and simply walked away. The only structure left was that drafty chicken coop, half filled with frozen manure.

We find it repugnant when people exploit or abuse others for personal gain—we call them cheats, tyrants, scoundrels, or villains; we describe them as despicable, evil, vile, wicked, or manipulative. Leopold said we should feel the same way about people who exploit or abuse land. If someone we meet is broken or damaged, we reach out to help them. If land is broken or damaged, we reach out to help it—by planting trees and native wildflowers.

For years after its publication in 1949, Leopold's little volume of essays had a tiny following. But today more than 2 million copies have been sold, and its ideas and ideals are considered a foundation of the conservation movement. At Tarboo Creek, Carl was watching his grandchildren carry on work his family had started seventy years before.

By the time the mussels and salmon fry were settled in and the cleanup accomplished, it was late afternoon. We gathered around Carl and watched the water flow by in the new channel, gently working its way toward the Pacific.

Trees

~~~

Among gardeners in Wisconsin and Minnesota, late January is famous for the arrival of the seed catalogs. The holidays are over; the temperature is below zero and dropping, and there is nowhere left to pile the snow that's still falling. Something like spring seems impossible. And then, precisely then, the seed catalogs arrive.

I feel the same way about placing our tree order each fall. Something as impossible as a forest seems just a hope away.

Our place along Tarboo Creek is shaped like a triangle, bordered by a one-lane road that runs southwest to northeast and a two-lane blacktopped road on the east. Biologically, the parcel has three distinct zones: a derelict pasture of about 6 acres that makes up the triangle's point, a 3-acre strip of floodplain on either side of the creek, and 8 to 9 acres of cutover, third-growth woodland on the bluff above the stream.

In 2004 the pasture was treeless. The floodplain had a single row of alder stems on either side of the ditch, along with a tangle of willow and alder in the remnants of the old creek channel. The uplands had been stripped of trees that were larger than a hand span across. We needed to reforest the old pasture and the floodplain, and underplant in the cutover uplands. As I write, we are twelve years and more than ten thousand trees and shrubs into the process.

~

Our tree-planting season can start in November—sometimes even October. Like the arrival of the salmon, the onset of planting season depends on the rains. If the winter storms are early and heavy and the soils re-wet sufficiently after the Pacific Northwest's late-summer drought, you can start putting trees into the ground before leaf drop is over. But if the rains are late and light, you would dull your shovel trying to dig at the bone-dry ground, then worry about the saplings' roots drying to death before they have a chance to acclimate and grow.

But whether you start early or late, the first thing to do is decide which trees and shrubs you're going to plant. This is actually harder than it sounds.

Traditionally, the goal of ecological restoration has been to replicate the conditions that existed when Native Americans were managing North America, before the arrival of Europeans. To do this, you go back to the original land survey records, when the government divided the country into a grid of 640-acre sections, and look at the notes the surveyors made about the vegetation on your site. In our case, Peter Bahls found references to "spruce bottom"—wet forests dominated by Sitka spruce.

The straight-grained, knot-free wood of old-growth Sitka spruce resonates so well that it is the material of choice for piano soundboards and guitar tops. Sitkas also helped persuade Bill Boeing to set up his airplane company in Seattle. In 1916, Sitka spruce was the preferred material for fuselage and wing frames; it's still the material of choice for building experimental aircraft. And it was abundant in the Pacific Northwest—especially in the cooler, wetter forests along the coast.

When the Salish people worked this land, before the arrival of whites, the spruce bottoms of Tarboo Creek and nearby Chimacum, Snow, and Salmon creeks were crisscrossed with beaver dams and spotted with ponds and marshes. They were wetland complexes—free-flowing streams that connected expanses of still waters ponded by beaver dams.

To supplement the information in the old land survey records, you can start walking. In our case, the destinations were stumps from old-growth trees that were felled in the 1860s and 1870s. When you find an intact stump, you're often able to locate notches for the spring-boards—planks the loggers nailed into the wide butts of gigantic trees. The springboards furnished a platform to swing double-bitted axes and pull a two-man crosscut saw—a tool they called the misery whip—so the tree could be cut above the flaring base.

Walking our place, we found some scattered western redcedar stumps, still 5 feet across at the cut after 130 or more years of decay. All of them were charred from the slash fires that had burned through the Tarboo Valley, and the rest of western Washington, between the years when the old growth came down and about 1920. So we added western redcedar to the planting list.

Another key destination is pockets of undisturbed vegetation. Back in the Midwest, where both Susan and I grew up and went to school, we'd walk railroad tracks to find little strips of intact prairie plant life. The railroads sometimes ran ahead of the farmers, so in a few places the tracks were laid and rights-of-way fenced before the prairies were plowed under—leaving relics we could use as a source of seeds and information about which plants grew at which sites. At Tarboo Creek, we walk the floodplain near the mouth of the stream, just north of its junction with the estuary and salt water. The tiny bit of old growth that remains there, near the tree we call the big Sitka, includes majestic, moss-covered bigleaf maples in the canopy, sprawling thickets of shrubby vine maples in the understory, and a forest floor covered with lush sprays of sword fern.

So in thinking about the future of our place, we could create a vision of the past from relicts and stumps. But by the time we were getting our restoration work under way, the science of climate change had become a sophisticated, maturing discipline. Research at the local and global level had made it clear that temperatures and precipitation patterns in western Washington were changing rapidly and would continue to change. The forests of 1820 would not thrive in the climate of 2120.

This makes things difficult. In some places on the Olympic Peninsula, western redcedars can live a thousand years. The big Sitka is at least five hundred years old, but it's growing at what is now the southern edge of its species' range. Do we plant trees for the past, or for the future?

~

Knowing which trees to plant requires an understanding of the nature of climate change. For me, this was a four-step process.

First, I looked at data on rates of fossil fuel use over time. The graph begins to swing upward at the start of the Industrial Revolution, then spikes as the industrialized nations moved to a petroleum-based economy after the Second World War. I also found a data set on the British Petroleum company's website documenting the average amount of oil used per person per year. The data were organized by country, so I made a map. It shows that northern countries tend to have high rates of fossil fuel use per person, with the United States having the highest of all.

Second, I found data on changes in the concentration of carbon dioxide, or CO_2, in Earth's atmosphere over time. The graph mirrors the pattern of fossil fuel use—steep increases since the Industrial Revolution that show no sign of slowing. The correlation is logical, because there is a causative link: when we burn gasoline or coal or fuel oil, we release CO_2 into the atmosphere.

Burning is what a chemist calls an oxidation reaction. In this case, the hydrocarbons and other organic compounds in fossil fuels react with oxygen and are transformed into carbon dioxide and water. The reaction releases the energy that keeps us warm in winter and powers our cars and trucks. A carefully controlled version of the same overall reaction is occurring in your body's cells right now, except that in your case, the organic compounds that are burning are sugars—not hydrocarbons. The energy that's released is used to power the chemical reactions keeping you alive, and the CO_2 that's released as a by-product returns to the atmosphere every time you exhale.

Trees and other organisms that perform photosynthesis oxidize organic compounds to stay alive just as we do, but they make them as well. They take in CO_2 molecules—possibly from your breath—and use the energy in sunlight to transform them into sugars. When we burn fossil fuels, then, we are burning the bodies of photosynthesizers and the creatures that ate them—releasing energy from sunlight that hit Earth hundreds of millions of years ago.

Third, I looked at a graph of average global temperature over the same period of time. Along with fossil fuel use, it is increasing. This made sense, too: CO_2 is particularly efficient at absorbing infrared radiation that's reflected from Earth's surface and is headed back into space. These infrared wavelengths are the same ones you've stood under if you have a heat lamp. When CO_2 absorbs infrared, the molecule gains energy and moves faster in response—meaning that its temperature goes up. As it bumps into nearby nitrogen (N_2), oxygen (O_2), and water molecules in the air, some of the heat that the CO_2 absorbed in the form of infrared radiation is transferred to the other molecules. As a result, all of the components that make up air begin moving faster and heating up the atmosphere.

Because heat that would otherwise be lost to space is being held in, Earth is getting warmer. Sea level is rising, too. In part this is due to the melting of alpine glaciers, the Greenland ice sheet, and polar ice. But most sea-level rise is actually caused by what oceanographers call thermal expansion: when you heat water, the individual molecules move faster and the water's volume increases.

Fourth, I looked at the projections for the future made by the Intergovernmental Panel on Climate Change, or IPCC—a Nobel Prize–winning group made up of the world's most experienced and respected climatologists. The latest (2013) IPCC report is based on computer models of how average annual temperature and precipitation, along with average sea level, will change around the world relative to the average values from 1986 to 2005. The panel did these projections based on four different scenarios about how much CO_2 we will put into the air during the upcoming decades—what they call representative concentration pathways, or RCPs. The lowest concentration pathway assumes two things: that future greenhouse gas emissions are as low as any value that's been used in the literature on climate change, and that they peak in the twenty-first century and are declining by the year 2100. This scenario is possible if stringent controls on emissions are implemented worldwide. In addition, there will have to be a dramatic slowdown in human population growth, a large-scale conversion from fossil fuels to renewable energy, and widespread implementation of carbon capture and storage technologies that don't yet exist. This Very Low scenario depends on achieving a 50-percent reduction in emissions from 2010 levels by 2050 and no net release of CO_2 by 2100. The highest RCP, in contrast, assumes that annual carbon emissions continue on their current upward trajectory, that human population approaches 12 billion by the year 2100, and that intensive use of fossil fuels—especially coal—continues. In effect, the High scenario is a present-trends-continue or business-as-usual projection. The other two pathways are intermediate between the Very Low and the High. Publishing four schemes, by the way, carries a message in itself: the climate that your grandchildren and great-grandchildren experience depends on what we do over the next few decades.

The IPCC represents the best expertise available, and to date their predictions have been correct. Their first reports with global predictions were published in 1990; a series of analyses published between 2007 and 2015 indicate that those initial projections, along with subsequent updates about changes in temperature and sea level rise, have been accurate. Thanks to the IPCC, we have a sound scientific understanding of how climate is likely to change in the future.

Finally, I asked: What does this mean for Tarboo Creek? In terms of changes in precipitation, the IPCC's models using middle-pathway greenhouse gas concentrations indicate that in 2100, winters in the Pacific Northwest will be about 8 percent wetter and summers about 8 percent drier, on average. This outcome would increase fire danger substantially, but otherwise most of the trees and shrubs on our planting list could probably cope. Temperature is a different story, though. Under the Very Low pathway, requiring extremely aggressive action on climate change, our average annual temperature in 2100 will be like the Oregon-Washington border today. Under the middle-road pathways, our corner of the world will feel like northern California; if the Very High, business-as-usual emissions pathway is the one we follow, average annual temperatures at our place will be what they are today in Silicon Valley. Our grandson, born in 2015, may live long enough to see that.

This is why things get difficult when it comes to ordering trees. The vegetation in the Bay Area is nothing like what we have near Tarboo Creek today. There is some overlap in northern California and extensive overlap in northern Oregon.

Recent research has already documented that the frequency and/or intensity of wildfires, droughts, hurricanes, and floods is increasing worldwide. In both the United States and on a global scale, the amount of money being spent on natural disaster relief and wildfire fighting, or lost in homes and commercial structures damaged in storms or fires, is rising dramatically—mimicking the increase in CO_2 and average global temperature. The data are like the sounds you'd hear if an elephant were moving around the room. Even if your eyes are closed, you know that something is going on.

If all this noise eventually wakes us up and inspires effective action, carbon dioxide emissions will slow, and we should plant the types of trees already found in western Washington—even though temperatures will probably continue to increase at least until 2050. But if we keep our hands clamped over our ears and sing loudly enough to ourselves, climate change will be severe enough that we would need to plant a forest like that of today's northern California or even Bay Area. The problem is that this isn't an option. The climate hasn't changed enough yet to let many of those species thrive on the Olympic Peninsula—even though it is projected to in fifty to a hundred years.

～

A recent study of British Columbia's forest types, which overlap extensively with the plant communities found in Washington, suggests that climate change will have two major impacts: today's high-elevation forest types and alpine meadows will shrink or disappear, and tree species that have a northern range limit in the province may find suitable climates advancing northward at up to 60 miles (100 km) per decade. The trees will have to hustle to keep up; it's not clear that all of them will make it.

For us, this research boils down to a series of do-we or don't-we questions. For example, do we plant incense cedar? This beautiful timber tree grows from the west coast of central Oregon down to northern Baja; the pencils you use are probably made from it. Local nurseries grow incense cedar in western Washington and distribute it with a bright yellow NON-NATIVE SPECIES tag. But as climate change continues, the range of incense cedar will head this way. It may be native to the area by the time our grandson is walking the banks of Tarboo Creek with his children. We could ask the same about Port Orford cedar—a tree that yields splendid wood for making arrows and currently grows in southern Oregon.

Another quandary: historically, restoration ecologists have been careful to replant with seeds or seedlings from nearby populations of native species to ensure that the individuals on the new site are similar, in terms of genetic makeup, to what grew there in the past. The idea was that local individuals are adapted to local environments, their genetics having been honed by centuries of evolution at that site. But with climate change, sourcing from local populations is less of a consideration; in some cases it may even be wise to get seeds and seedlings from sites farther south or at lower elevation.

What to do? We are optimists. In the 1950s, Seattle's Lake Washington was so polluted by a daily dose of 20 million gallons of poorly treated sewage that children weren't allowed to swim there. There was a general outcry—an expression of what Bill Gates Sr. likes to call the public will. In response, a countywide agency was formed to build and operate a comprehensive municipal sewer system. Water quality in the lake improved quickly, and our boys swam in it as kids. One even did a New Year's Day polar-bear plunge: he came out cold but clean. Similar scenes occurred throughout North America in the 1950s and '60s, and the widespread eutrophication of lakes slowed or stopped. In freshwater systems, death through overfertilization is largely a thing of the past.

In the 1970s and '80s, extensive coal burning to generate electricity in North America and Europe, along with increasing gasoline consumption by cars and trucks, was introducing enormous quantities of sulfur oxides and nitrogen oxides into the air. When these compounds react with water in the atmosphere, the products are sulfuric acid, nitric acid, and nitrous acid—the active ingredients in acid rain. In northern Europe and the northeastern United States, statues were disintegrating and forests were dying. In response, there was another expression of the public will. The U.S. Clean Air Act Amendments of 1990, and similar legislation in other industrialized countries, started a slow but steady decline in the amount of sulfur oxides and nitrogen oxides released into the atmosphere. Forests and lakes began recovering. The decline in acid rain is still continuing, even though electricity generation and economic output have grown dramatically.

In the 1980s through the early '90s, the percentage of ozone in the upper atmosphere began decreasing; an ozone hole opened over Antarctica each Southern Hemisphere summer. Ozone absorbs the ultraviolet wavelengths in sunlight and thus protects Earth's surface from the highest-energy, most dangerous components of sunshine. It's a sunscreen. The ozone layer was thinning largely because chlorine atoms, off-gassed from refrigerators and air conditioners, were triggering chemical reactions that break ozone molecules apart. Less ozone means more skin cancer and photoaging of skin. In response to the data, there was a third expression of the public will: the international community signed the Montreal Protocol in 1987, phasing out the production of ozone-depleting substances. Limitations on these substances have started to produce significant results. There is still much to be done, but the percentage of ozone in the atmosphere is rebounding. The Southern Hemisphere's ozone hole has stabilized and may even be starting to shrink.

Similarly, partnerships between government agencies, businesses, and nonprofit organizations in the past fifty years have led to dramatic reductions in littering and smoking. The public will is powerful.

We are optimistic that another expression of the public will—one that starts a slow but steady decline in carbon dioxide emissions—is beginning. Like the response to acid rain and ozone depletion, the response to climate change will have to be international and long term. The Kyoto Protocol, negotiated in 1997, set targets for decreases in greenhouse gas emissions and was signed by virtually every country in the world except the United States. Changes in behavior may be particularly wrenching for Americans because the country developed in an era of cheap gasoline and subsidized highways. The result is land-use patterns that make energy-efficient transportation difficult—I know someone who drives 130 miles round-trip to work each day. So resistance to change is high. But change is also opportunity: people who can adapt and innovate will thrive, while people who look backward and maintain a lifestyle dependent on fossil fuels will see a larger and larger percentage of their income disappear in exhaust. And one of the great messages of my adult life is this: never underestimate the energy, resolve, and creativity of the American people. Indeed, an impressive array of states and municipalities have set ambitious targets for reducing greenhouse gas emissions and limiting future temperature increases—targets that move beyond the terms of the Kyoto Protocol.

So at Tarboo Creek, we plant trees that will grow well now, and we work toward a future that doesn't depend on fossil fuels. If that future doesn't arrive, our great-grandchildren will have to replant.

~

So back to the question: What to order?

We get most of our plants from state or county agencies that maintain nurseries and sell to private individuals or organizations, but
on occasion we grow some of our own. Madrone trees, for example,
are difficult to get from the nurseries, so our older son started a little
bank of seedlings from fruit that fell on sidewalks near our home in
Seattle. Madrones are a broad-leaved evergreen tree, an anomaly in
northern latitudes. They have shaggy orange and brown bark and
dominate bluffs and other gravelly sites—especially near the coast—
where other trees struggle. We found a few in our woods when we
first bought the place so wanted to increase their numbers. We've
also worked on growing Sitka mountain ash, beaked hazelnut, and
evergreen huckleberry.

Almost everything else we need, though, is available from sup-
pliers. So in fall we sit down, look over the websites published by
the nurseries, and figure out how many individuals we want of each
species on their lists. We do this figuring based on the old gardener's
dictum: put the right plant in the right place. To grow a forest well,
you have to know your site and your trees.

As we walk over an area we're going to plant, we try to notice the
exposure and the slope. How much sun will the trees get here, and
how quickly will water move downslope? Sitka spruce love to have
their feet wet, and western redcedar can form lush, single-species
stands in wet, boggy spots with blackened, peaty soils. It's not uncommon to find bigleaf maple on wet sites like creek bottoms, but both
bigleaf maple and western redcedar also grow well upslope, out of
the wet. Red alder is a pioneer and relatively short-lived—it invades
abandoned roads and clear-cuts and can form stands as dense as
dog hair. It partners with nitrogen-fixing bacteria that live inside its
root cells so has a ready supply of fertilizer in nutrient-poor sites.
Douglas-fir, grand fir, and western white pine tend to be generalists

that can thrive on an array of sites away from creek bottoms and wetlands.

Designing a planting is a little like arranging places at a holiday dinner with family and friends. Everyone has their quirks and strengths and weaknesses; you need to know them and place people accordingly. For example, we've been planting red alders close to Sitka spruce ever since one of our boys found a study showing that once alders age and begin to die back, Sitkas get a huge shot in the roots from the nitrogen in the soil.

But we've been fooled, too. The 6 acres of gently sloping land on the east side of Tarboo Creek was pastured for a hundred years and left empty for thirty. It was an upland, sunny location with a southwest exposure. So we planted a mix of trees dominated by Douglas-fir, grand fir, and western redcedar. But in 2 acres of this area, in the northernmost tip of the property, virtually nothing survived.

At first we thought the problem was the crew of local schoolchildren and families that had done much of the planting. Using kid power was the brainchild of Northwest Watershed Institute, who realized that plant-a-thons would be not only a great community event but also potentially a fundraiser for education. NWI staff recruited two local grade schools, and the kids started selling cards promising that they would plant a tree in return for a small donation. In recent years as many as four schools and five hundred students and parents have been involved in a single planting season. Plant-a-thons have now reforested much of the Tarboo Creek floodplain and touched lives in unforeseen ways. When our dog Cam was dying of stomach cancer and got so sick she couldn't stand anymore, a veterinarian in nearby Port Hadlock kindly put her down. The vet had no idea we were involved in the restoration project, but a week later we got a card from her office saying a tree had been planted in Cam's memory at a plant-a-thon near Tarboo Creek. I've also heard college students I didn't know give speeches saying that their interest in the natural world started with a Tarboo Creek plant-a-thon they participated in as schoolchildren.

Our site was the first-ever plant-a-thon that NWI organized, though, and the kids and parents were new to planting. Now they are old pros—some have planted every year for ten years or more—but back then, we feared operator error in the patch where all the trees had died.

Failure like this isn't all that unusual. The first season that the Leopold family planted pines on their property in Wisconsin, the spring of 1936, one of the Dust Bowl–era droughts killed everything—three thousand saplings. The family did what good people do in a situation like that: they replanted.

Our little 2-acre tree cemetery became known as the Dead Zone. We replanted the next year, and the year after that. Still no success. By now it was clear that it wasn't operator error at all—the plant-a-thoners had done just fine. Something else was going on.

Finally, late one winter, I figured it out. I was trying to find a few saplings among the hundreds of dead bones sticking up when I noticed that the bases of the needleless trees were standing in water. The site was upslope but sopping wet. I recalled having an easy time digging holes for new saplings in the area because the soil was so fine and homogenous. When I also remembered that the soil was sticky, a light finally went on. It was clay.

Clay consists of particles that are a fraction of the size of sand grains. Because they are small and light, they stay suspended in moving water long after sands and gravels fall out. Clay particles are routinely carried long distances by rainwater or glacial melt before being deposited in low spots when the water finally slows down. Many of the five-story-high bluffs above Puget Sound consist of clay, deposited at the bottom of a lake that formed during glacial times.

The Dead Zone started just 30 yards from the creek bottom and was on a west-facing 10-percent slope. But for some reason, there was a clay lens close to the surface. Because clay particles pack so tightly that there is almost no space between them and water can't percolate through, the clay layer was acting like a swimming pool liner, creating what hydrologists call a perched wetland. The Dead Zone was dead because the trees were drowning.

A few years later, the biologists in charge of restoring the Elwha River floodplain in Washington began facing the same problem, writ extremely large. The Elwha has headwaters in the center of the Olympic Mountains and flows north to the Strait of Juan de Fuca. (The Olympics are among the only mountain ranges in the world with radial drainage, meaning rivers flow out from them in all directions.) In 2011 the National Park Service began removing two dams across the Elwha in what qualifies as the largest dam removal project in history. The hope is that a restored river might again host runs of Chinook salmon with individuals as large as those recorded in pre-dam days: 100 pounds.

When the dams were down and the lakes behind them had drained into a free-flowing river channel, biologists were left with hundreds of acres of bare clay—formed from sediments that used to be carried down to the Pacific but that had settled out of the quiet lake water. Some parts of the clay beds were dotted with old-growth stumps, but the original forest soils were buried. The clay would be waterlogged in winter and baked as hard as pottery in summer. The plantings were designed appropriately but had to contend with yet another issue: elk. The Olympic Peninsula is home to a subspecies called Roosevelt elk, and when the Elwha restoration was newly planted the

elk seemed to take special pleasure in going from sapling to sapling, pulling them up or browsing them down.

Although young male elk sometimes wander through the Tarboo watershed after they've been weaned, we don't have anything like the resident herd that the Elwha managers have to contend with. The Dead Zone has been a challenge, but we consider ourselves lucky in comparison.

So our tree order changes each year as we learn more about our soils and site, and as saplings planted in earlier years thrive or die. The Dead Zone doesn't fool us anymore—when we're planning for that area now, we order things like Sitka spruce instead of the usual array of upland species. We've gone so far as to try western cotton-wood—a classic river-bottom tree—and willows. And when we order for areas where the early plantings of trees are doing well, we load up on shrubs so we can get an understory layer started. The only constant is change. Each year we watch, learn from the land, and try new things.

～

Tree orders get mailed off in October, but we usually wait to take delivery until early January. By then the holidays are over, temperatures are cool and steady, and the soils are well wetted.

Between ordering time and planting time, then, we have a couple of months to wait. And as we're waiting, the fish come.

Salmon

~~~~~~~~~

There is poetry in a salmon run. When a big year is peaking, the creek is a frenzy of flashing red skin, fighting, splashing, and sex. It is a bacchanal, a primal rite passed down from the ancients. But when a run fails, there is a void. The landscape is changed, for its soul is missing. You stare at the water flowing by, wondering if its life, and your life, will always be this empty.

There is mystery in a salmon run, too. These are large creatures; even in tiny streams like Tarboo Creek, many are 2 feet long and 7 pounds or more. But they appear from nowhere. Days and weeks and months go quietly by; then suddenly the stream is alive with great beasts, thrashing their way upstream. And there are so many questions: Why are they abundant enough in some years to crowd each other, and almost nonexistent in others? When you see them migrating in the open water of Puget Sound, before they sort themselves out by natal stream and thus final destination, they throw themselves up and out of the water: one, then another, flashing like silver coins flipped in sunshine. Why do they do this? Then how do they find their way to their fryhood home and adjust to living in freshwater, after years in the salt?

Finally, there is beauty in a salmon run. When the fish approach the stream of their birth, their bodies change. The males acquire great hooked jaws—cartilaginous extensions to their normal bony jaws—lined with triangular teeth that are useless for hunting prey but efficient in combat. They have morphed into fighting machines, brandishing spiked war clubs that they wield in fights over females. Both males and females color, their silver bodies darkening as salmon-colored pigment molecules migrate from their muscular sides to concentrate in their skin. Females also load these pigments into their eggs.

The orange-red pigments that color breeding salmon and their eggs are members of a molecular family called the carotenoids; they make egg yolks an orangey yellow and northern cardinals a bright red. Carotenoids are synthesized by photosynthetic organisms—the plants and algae and bacteria that do most of the planet's interesting biochemistry—and are passed up the food chain to herbivores and their predators. Because the pigments are acquired in the diet, only well-fed individuals can have good carotenoid supplies. The pigments boost immune system function in both birds and fish—individuals subsisting on a carotenoid-poor diet are not only more dully colored but also tend to be sicker. So colorful salmon are healthy, well-fed salmon. Both males and females actively prefer redder mates.

The sockeye and pink salmon found in larger streams in the Pacific Northwest flush scarlet from nose to tail when they move into rivers to breed. Tarboo Creek's coho dress in a deep emerald green above and a brilliant red below; the chum salmon are olive streaked with ruby-red gashes, like lightning strikes.

Along with changing their jaws and body color before spawning, salmon remake their gonads. The males load long tubes with milky-white milt, brimming with tens of millions of sperm, and the females ripen great sacs of eggs, numbering up to seventy-five hundred in large species like Chinook. The load typically represents a fifth of a female's body weight; imagine a 140-pound woman carrying a 28-pound baby in utero and swimming upstream—in some cases, hundreds of miles; in many cases, through boulder-filled rapids and up and over waterfalls.

The salmon of Tarboo Creek are large, sleek, and strong, but ephemeral. They arrive a fire-engine red and bristling with vigor but are dead in less than two weeks. The females beat the skin and muscle right off of their tails as they dig a nest for their eggs; it's common to see exposed bone in the tails of spawned-out coho. Even before spawning is over, a mother's eyes may begin to dim. Soon they whiten and begin staring out at the world haunted and ghostlike. Her brilliant breeding coloration fades quickly, and patches of infected skin appear—fuzzy with growing white strands of the funguslike decomposers called oomycetes. Her body has started to rot before death. Within days, you will find her carcass lying on the shallows, often next to her nest. She has given everything to the young. Nearby, dying males will lie on sandbars, slowly being covered with sediment, even as their gills still open and shut. Brief floods may carry the bodies away a short distance, only to be snagged on the branches of low-lying trees, where they hang, limp. Newly arrived salmon, fresh and brightly colored, swim by the carcasses—passing the dead on their way to the front.

~

At the start of a run, when they are moving in storm-driven water, the salmon are quicksilver. If you get too close to the stream they flash away; all you see are the ripples of their splash. The fish themselves have disappeared into the shadows or raced upstream—heading to a promised land of deep gravel and many mates. But if you are lucky enough to spot a female who has already committed hours to excavating a particular site, she will not flee. She will ignore you. So will the males who have come to love her, biding their time nearby. If you sit quietly, you can watch a nest taking shape, at your leisure.

We did this, one afternoon on a Thanksgiving weekend. Susan found a female digging a redd in the gravel directly above a riffle, where the water was bubbling rapidly over a cobbled incline. This is a prime location for salmon eggs to develop: the gravel is deep and well oxygenated. It also happened to be in one of the first meanders Tom and Peter had engineered and Bob had dug. The female was nesting in newly restored habitat.

Susan was joined by our boys and a cousin; some friends and I came and went throughout the afternoon as we did stream-maintenance chores or walked other sections of creek looking for redds. The new nest was in a section of the stream that wasn't more than 6 or 7 feet across; the water was shallow enough to see the bottom clearly. Sitting on the bank with the toes of their boots in the water, Susan and the boys could almost reach out and touch the fish.

Susan named the female Cassandra, inspired by her sleek, exotic looks and—once we'd counted the courting males nearby—her powerful allure.

When we first started the watch, there were three males in attendance. Closest to Cassandra—currently in pole position—was a beautiful hunk of a male, clearly fresh from the ocean. The boys called him Fabio, to reflect his classic European styling and flair. He was the type of big, flamboyant fish that makes all the girls stop and stare. Nearby was Tony Bennett, who may have achieved greatness in the past but by now had seen better days. His coloration was a shadow of its former self; he was in the game, but out. The third gentleman-in-waiting was Red-cheeks. When the males jostled for position near Cassandra, Tony Bennett was sometimes displaced by the fresher Red-cheeks and always by Fabio.

Although it got difficult to keep track, Susan eventually counted at least six males who were following Cassandra's progress. Fabio, Red-cheeks, and Tony Bennett tended to stay nearby; the other

salmon-boys would check in for a few minutes and then head back upstream—presumably to other riffles where other females were preparing their nuptial beds. Those males were making the rounds, keeping tabs on the action. They had to weigh their prospects with various females, assessing their odds against the other suitors at each redd. But sooner or later, they all came back to Cassandra.

As the afternoon wore on, Susan noticed a fourth fish staying close to the redd. He was small and secretive, staying well away from Fabio, Red-cheeks, and the other big boys. He was probably a jack—a sneaker. Jacks occur in several salmon species, including coho. They are males that only stay out in the ocean for a single year instead of the multiyear residence—two or three or four, depending on the species—of normal males. Because jacks are younger, they don't get big enough to compete directly for females. Instead, they loiter on the street corners near redds, waiting. When a female finally begins to spawn and the dominant male aligns with her, the jack will dash to the female's opposite side, spray sperm at the eggs, and bolt for cover before the attending male can bite him.

Jacks can play a winning hand if two conditions are met. First, predation pressure in the ocean, from orcas and fishermen, has to be intense—making life dangerous for males who stay out multiple years. The big boys may get bigger each year, but they may also get eaten. The key insight here is that jacks are making a trade-off between higher survival and lower success at fertilizing eggs. They are less likely to be eaten by orcas, but they also can't hope to best the big males in fights over females. So for the jack way of life to pencil out, a lot of salmon have to get eaten in their second (or third or fourth) year out in the ocean. But there is another issue: other jacks have to be relatively rare. If the creek is full of sneakers, the jacks start competing with each other as well as the big boys. The dominant males also get more wary of them—meaning that jacks have a greater chance of getting killed in flagrante delicto by a jealous husband.

Breeding experiments have shown that the "decision" to be a jack is at least partially determined by an individual's genetic makeup. That is, some males have a predisposition to leave the ocean early and pursue what biologists like to call an alternative mating strategy. The frequency of these sneaker alleles—an allele is a version of a gene—goes up and down over time, as jacks do better or worse in the game of fertilizing eggs.

~

All of these little dramas involving males, though, were nothing compared to what Cassandra was going through. We would watch her float above the redd placidly for five minutes or more, then suddenly roll to one side, arch herself into a U, and beat her tail like a hoe against the rocks—making four or five quick, forceful thrusts. It was over in a flash and a splash. Then she would float again, peacefully. It was like watching an Olympic champion still himself, clean and jerk two huge iron disks with a great roar of effort, drop the barbell with a crash, and then stand there, dazed and rocking slightly. The males—who swam aside each time Cassandra pitched herself into a bout of digging—would sidle back toward her.

At first we wondered how long she could keep this up. But then we realized that the answer was simple: as long as it takes. The depression Cassandra was carving had to be deep enough for her eggs to settle into—out of the current—until they were fertilized and she could cover them back up. It also had to be deep enough that the eggs wouldn't be exposed by scouring currents—to prevent them from being swept downstream when winter storms brought the creek to flood stage. But too deep, and the tiny fry, still carrying their yolk sacs, would have trouble wriggling up to the surface after hatching. Redd making is a Goldilocks problem: not too deep, not too shallow—juuust right.

Susan and the boys spent most of the afternoon by the reddside, sitting shoulder to shoulder. They marked Cassandra's progress, commented on the comings and goings of the salmon boys, listened for the bald eagle calling downstream, and watched flocks of chickadees and kinglets as they went foraging by.

And always, they watched the water. A creek is flowing rain: water that rose into the air somewhere in the wilds of the North Pacific before being transported south by a swirling storm front that smashed into the Olympic Mountains and dropped its load of moisture. The water in the creek right now could be fresh drippings from the forests of the upper watershed, or fossils that had been sitting in the ground for years, slowly seeping their way downhill.

The salmon also follow this cycle, from ocean to forest to river and back. As they move through the open ocean, the growing fish store nutrients that have welled up from the depths of the sea. The nitrogen, potassium, and calcium atoms in their bodies were passed from photosynthetic bacteria to microscopic crustaceans to herring and candlefish to them. When they've grown stout enough to carry the load, the salmon swim these nutrients back to the land. As they spawn and die, the nutrients may be snatched up by the bacteria that coat the gravels and fill the muds in the creek, only to be passed to a caddis fly nymph and a tiny salmon fry. In this way, the young fish are nourished by the bodies of their mothers and fathers. Or the parents' bodies can be hauled away from the creek by a river otter, bald eagle, bear, cougar, raccoon, or raven. The scavenger's feces, along with the remains of the salmon carcass, fertilize the trees. In this way, the forest is fed by the fish.

Experiments that trace the path of nitrogen atoms in forests have documented that a tree will quickly take up more than a third of the

nitrogen in a salmon carcass deposited under it; an observational study in southeast Alaska showed that Sitka spruce near salmon streams grow three times as fast as Sitkas on similar but salmon-less sites nearby. The 20 million sockeye salmon that run into rivers feeding Alaska's Bristol Bay can deposit 54,000,000 kg of biomass in the adjacent watersheds. Salmon are a swimming fertilizer. They represent a massive transfer of nutrients from the open ocean back onto land.

Eventually, the Sitka spruce and other trees along salmon streams drop their needles and leaves back into the creek, where the leaching nutrients are absorbed by bacteria and algae and passed back up the aquatic food chain. One way or another, the atoms that make up the older generation are passed on to the new.

~

If you watch a creek flow by long enough, you will find yourself think-ing about things: like how long this cycle of water and salmon has been going on without us, or how much you love the people sitting next to you, or how long it's been since you just sat somewhere. And you begin to wonder about things: like what this spot was like during the last ice age, when there was a mile of ice above you, and whether fish feel weather. A salmon can't possibly understand humidity or wind or rain or snow. It can only know that the water around it is cold or warm, fast or slow, deep or shallow, fresh or salty.

But salmon can feel things we can't. Along with other fish and aquatic amphibians, they have a lateral line—a sensing system that runs the length of their bodies on either side. The core of the system is a row of sensory cells similar to the ones found in your inner ear. These cells have tiny hairlike projections that bend in response to changes in pressure. The bending movement causes a minuscule electrical current to flow inside the sensory cell, which triggers a stronger electrical signal from adjacent nerve cells to the brain. The brain integrates the signals sent by all of the cells in the system and sends a pulse of messages back down motor neurons to the muscles that control swimming. In response, salmon face into water currents. They detect rocks and other objects nearby and avoid smashing into them; they sense and turn toward passing prey.

Salmon feel their way, and not only with the lateral line system. The leading hypothesis for how they navigate hundreds of miles of open ocean back to the region of their natal stream is magnetosensing. The idea is that they construct a map based on perceiving and recording changes in the intensity and angle of Earth's magnetic field. It's an ability that humans can only imagine—or replicate with sensitive instruments.

Once the fish have followed this magnetic map back to their old neighborhood, they use their sophisticated olfactory bulbs to smell their way home. It's like my parents navigating to my father's hometown in Indiana using a highway map and then following landmarks to get to my grandparents' house. My grandmother always baked cookies in anticipation of our arrival; my brothers and sisters and I would jump out of the car and home to her kitchen, following the smell.

In the Columbia River, some salmon follow their noses like this for 600 miles, making an elevation gain of 2,000 feet, to get to the same stretch of the same tributary where their parents spawned. It's hard to exaggerate just how specific this homing can be. Tagging experiments in southwest Alaska have shown that salmon eggs buried in lake gravel, versus the gravel in nearby streams, give rise to adults that return to the lake—not the streams a stone's throw away.

A salmon's sense of smell does much more than guide it, though. Salmon can smell danger. Experiments have shown that they avoid water that has been washed across the skins of bears or otters. They can smell food nearby, too—both as fry in streams and as adults in the open ocean.

Salmon also see the world differently from us. With eyes positioned on either side of their head, they get a much wider perspective on things than we can. The trade-off is that they lack some of the depth perception we're capable of. Further, inside the receptor cells that line their retinas, color-detecting pigments can monitor the angle at which light is scattered or polarized—another sense we lack. These eye pigments are also particularly good at distinguishing wavelengths of light in the ultraviolet (UV) and red part of the spectrum. The red sensitivity is probably an evolutionary response to the importance of red coloration in mating and other aspects of salmon social life; the best hypothesis we have to explain the UV sensitivity focuses on seeing well enough to feed near the water surface. As sunburn victims know all too well, bright sunlight is packed with ultraviolet wavelengths; UV is also abundant in the first few centimeters below a water surface. Pigment molecules that detect UV are present in the eyes of newly hatched fry, which feed at or near the stream surface, and are thought to help the fry feed and grow efficiently in freshwater. The UV-detecting molecules are lost, however, when juvenile salmon undergo smoltification—the drastic physiological remodeling that allows a freshwater fish to move into salt water and thrive. Instead, young salmon that are making the freshwater-to-salt-water transition acquire an extra sensitivity to blue wavelengths, which are abundant in the deeper ocean water they'll soon call home. The ability to detect UV is regained when adults return to freshwater to breed—a change that helps their eyes cope with the light available in shallow water. It's no wonder that Darwin referred to the vertebrate eye as an organ of extreme perfection.

The human eye, in contrast, can adjust to differences in light intensity, but the types and distribution of pigments in our eyes don't change

as we descend into a darkened basement or even as we age. And we don't experience anything like the other changes that occur during smoltification. As a fish is moving from freshwater to salt water, its gills and its excretory system have to be completely remodeled. The water and the salts in a solution each move from areas where they are at high concentration to areas of low concentration. In a lake or stream, water tends to move into a fish's body across the gills and gut wall, and salts tend to move out. To cope, freshwater fish don't drink. They also have to rid themselves of excess water by urinating large volumes, and their gills and gut are constantly pumping salts back into the body. But in the ocean, the situation flips: water tends to move out of the body and salts tend to move in. Marine fish have to drink to avoid dehydration and then pump salts out like mad to avoid electrolyte poisoning. When a young salmon leaves its home stream to head out into the wide world of the ocean, it has to be prepared.

~

When we returned to Cassandra's redd the next morning, she was gone. There was no sign of her, save for a gleaming mound of fresh gravel. This is how you spot redds as you walk along a salmon stream: by looking for clean mounds of gravel, especially near riffles. The freshly turned stones contrast with the unexcavated rocks nearby, which have been darkened by a year or more of bacterial and algal growth. Cassandra had worked through the night and finished her magnum opus.

Peter Bahls thinks that the coho and chum females in Tarboo Creek build a single nest, but research has shown that in some species and river systems, a female may build four or five redds with five hundred to a thousand eggs in each before she is spawned out. This represents a lot of digging. In streams with big runs of 60- and 70-pound Chinook salmon, the spawning females actually rework the streambed—collectively moving hundreds of tons of gravel. It's like a giant sandbox filled with girls playing earnestly with toy excavators.

A biologist used historical records and recent data to conclude that on average, salmon in southwest Alaskan rivers dig up at least 30 percent of the total streambed each year. As they till the substrate, small particles are swept downstream and coarser pebbles and cobbles are left behind. As a result, the fish become part of the forces determining the stream's shape and course, along with the floodwaters that carve banks and excavate new channels, the fallen trees that deflect the current and create swirling eddies, and the logjams that dam flows and create deep pools. Salmon are movers and shakers. They are ecosystem engineers.

We never did find Cassandra's carcass, or the remains of Fabio, Tony Bennett, Red-cheeks, or the other salmon boys. But days later, Sean Gallagher fastened a length of yellow surveyor's tape on an alder twig near Cassandra's redd and marked it "Coho 11/28."

Sometimes I thought of it as a grave marker; at other times as the plastic bracelet around a newborn's wrist.

Peter Bahls and other NWI staff do spawning surveys every fall—following the fish upstream much as the bald eagles and river otters do, tying yellow or orange streamers to twigs near redds, and recording the locations on a map. When the creek was first remeandered at our place and fresh gravel was everywhere, we had thirteen of those yellow streamers on our 1,500 feet of channel. We were thrilled, but it was nothing compared to what we've seen on a tributary of the Hoh River, in the pristine rainforests of Washington's Pacific coast, when the Chinook are spawning. Biologists from Olympic National Park had been marking redds for weeks with yellow surveyor's tape before one of our visits, and the nests were so dense that the tangle of vine maple branches looked like a Christmas tree covered with tinsel. And there were still fish everywhere: females digging new redds virtually on top of old redds; males chasing each other and being chased—by two river otters. The otters were so full of fish that their bellies bulged, but they couldn't help swimming after another Chinook every few minutes. We saw one big male salmon fight back: he rammed one of the otters in the side and chased it out of the pool where his lady love was digging.

⁓

Now that the creek at our place is a few years older and has had a chance to remake itself, we have fewer good sites for redds. Much of the gravel is buried in silt and organic debris now; some has been piled into ridges or transported downstream by floods. The gradient through our land is relatively gentle, so the water moves slowly enough that it provides better habitat for rearing than nesting. Most of the adult coho swim through our stretch of creek now, headed for a smaller, faster, and gravel-lined stretch of stream closer to the headwaters. But come March and April, we start to see coho fry in the slower, mud-lined, and shaded portions of our creek. They've floated down from the redd sites and will spend their days eating and avoiding being eaten. Their time in the newly remeandered stream will fuel the growth that allows them to head to the ocean about the time we're celebrating their first birthday.

In the summer, we can sit at the edge of the stream and watch these small fry feed. The best place for this babysitting is a rock that Bob put on a creek corner. It's where the new watercourse leaves the old ditch—just upstream from the spot where our sand-bag dam diverted the flow. This little patch of creek is ideal rearing habitat because the water slows down before it hits the earth dam that Bob made to fill the ditch. And just below the big rock where we sit, the water fans out into a broad, shallow channel that Bob scraped out using an I-beam clenched between the teeth of the excavator's bucket.

The water in this stretch of stream is shaded by foot-wide alders that used to line the ditch and now overhang the creek. The leaves and twigs they've dropped into the water have created a velvety lining of rich organic debris. The big logs that Bob poked in have trapped sediments brought in by winter floods and created down-stream pools; the willow stakes that our boys and Sean and I pounded in have sprouted into 15-foot-tall bouquets.

So on an August afternoon, you can perch on Bob's rock just above the water or straddle one of the big logs he angled into the pool upstream. Either way, the creek is flowing under and around you. In the early evening, the sunlight dances through the moving alder branches and turns the creek a golden brown. If you sit still long enough, the little coho emerge one by one from the shadows under the big logs. They start fanning their tails to maintain position in the gentle current, then dart here and there to pick an insect or bit of debris from the surface. After fifteen or twenty minutes of this, you stand up and continue your walk—feeling much better, thank you.

~

Chum salmon fry move into the ocean almost immediately after hatching in March or April, but the coho babies stay in the creek all summer, not migrating down to the sea until just before the next generation of parents returns in late fall. Both species spend two years in the ocean, feeding and growing, and then return to Tarboo Creek to spawn. They're resplendent in their nuptial coloration, and they're big. Jeff Delia, an oyster grower who does catch-and-release fly casting when the salmon are schooling in Tarboo Bay, has hooked fish that were more than 2 feet long and weighed 15 pounds.

It's natural to wonder why salmon move from salt water to freshwater to breed. The leading explanation for this anadromous lifestyle focuses on two factors: food and predation. Freshwater is a much safer place for a salmon egg or fry than the ocean—there are many fewer midsize fish around to eat them. At Tarboo, almost nothing digs up the eggs, and the major threats to fry are kingfishers, great blue herons, and cutthroat trout. So freshwater is a good place for a juvenile fish to live during the most vulnerable period of its life. But streams, lakes, and ponds don't offer nearly as much food as the ocean does, making salt water a better place for young fish to grow to adult size. The populations of sockeye salmon called kokanee illustrate this second point. Kokanee don't migrate to the ocean; instead, they stay in freshwater year-round. They never attain the size of the oceangoing sockeye and can't compete with them for the best nesting sites. Like the jacks that appear in many salmon species, kokanee are making a trade-off between higher survival and lower reproduction.

People also want to know why salmon breed only once before dying. It's not that the to-and-fro life makes repeated breeding impossible: sturgeon spend most of the year in salt water and breed in freshwater, but they reproduce every few years of their adult lives and can outlive a human. Salmon appear to do better as big-bang reproducers because their odds of surviving to make another long-distance trip back to their natal stream, after a return to the ocean, are low. In addition, females that throw all of their resources into one reproductive event can produce larger eggs—which give their fry an important survival advantage—and more of them. Apparently, evolution has favored individuals that put all of their effort into a single bout of reproduction over fish that hold back but then lay fewer and smaller eggs and risk dying before the next breeding season arrives.

To a biologist, salmon offer a treasure trove of questions, and just enough answers to make the research effort rewarding. But even seasoned salmon researchers never fail to thrill at the sight of a spawning run in full swing—still feeling the excitement of that little boy in an upstairs bedroom near Tarboo Creek decades ago, listening to the sound of splashing in the night.

⁓

Peter Bahls, Sean Gallagher, and other NWI staff and volunteers have been counting salmon in Tarboo Creek since 2002. In addition to mapping and marking each redd in spawning surveys conducted during the fall, they sometimes livetrap and tally juvenile coho during the summer.

The numbers go up and down over time. I've worked near the creek on days when I heard a coho splash by every five or ten minutes. I've seen hordes of chum, near the big Sitka spruce tree at the mouth of the creek, churn the water like an outboard motor— spawning in a frenzied tangle of bodies. We've also had autumns when we never see a fish and can find just one or two pink stream-ers marking redds in the upper portion of the watershed. They dangle in the wind like two lonely people in a hall that should be filled with revelers on New Year's Eve.

Sometimes we don't know why the numbers are low; sometimes we do. The coho and chum that spawn in Tarboo Creek usually come into Tarboo Bay from the open ocean by late September or early October. Two large sand spits separate Tarboo Bay from Dabob Bay and the broader waters of Hood Canal. Once the salmon swim around the spits and enter Tarboo Bay, they are virtually trapped there—constricted in terms of how far they can roam. They wait in the bay, biding their time until the fall rains arrive to swell the creek and trigger the run. The situation is good for salmon watching: on a bright day you can sit by the water and watch them break the stillness with flashing jumps, or try to count their shad-owy silhouettes if they swim close to the shore. But we've also seen harbor seals take advantage of the geography by feeding on salmon penned in the bay. And one fall an NWI volunteer saw a fishing boat pull up, anchor a net on the beach, tow the line around the schooling fish, and scoop them up in a couple of sets. The crew came back the next day and did the same thing. The following year, it happened again. For two consecutive autumns, almost no fish made it up the creek to breed.

The NWI observer who watched the beach sets claimed that the single boat took more than six hundred fish in that two-day span. A neighbor who spent twenty-five seasons fishing for sockeye salmon in Bristol Bay, Alaska, estimates that the two-day catch was worth $3,600 at the most. In the whole scheme of things, this is a tiny amount of money. It's nothing compared to the time and effort government agencies, tribes, and private citizens have spent restoring habitat in the Tarboo watershed.

One of the salient issues in play is that Hood Canal salmon are managed as an aggregate—meaning that managers don't care if the salmon spawning in a particular stream are wiped out as long as the total number of fish returning to the region remains fairly constant. In addition, the state runs a hatchery for coho in Quilcene Bay, just to the west of Tarboo Bay, and some of the managers insist that the fish being taken in Tarboo Bay are—all evidence to the contrary— actually hatchery-bred individuals from the Quilcene River facility and not wild fish. There are virtually no restrictions on taking hatchery-raised salmon.

Because Tarboo Creek is so small, the extinction of its chum and coho wouldn't register in the regional totals. According to the way the regulations for managing salmon in Hood Canal are currently written, it would be perfectly legal to fish the Tarboo Creek populations to extinction.

~

Fishing salmon to extirpation is a story that started centuries ago; it was chronicled by scientist David Montgomery in a 2003 book titled *King of Fish*. Atlantic salmon, for example, were once abundant throughout Europe. The Roman legions ate salmon caught in downtown London; an 83-pound individual was harvested from the River Thames in 1821. A quarter million salmon used to swim up the Rhine each year, with some reaching 50 pounds. French rivers once teemed with them. Now, salmon are virtually extinct from Spain to Estonia. Only Iceland's runs are holding up reasonably well.

Unlike the situation on Tarboo Creek, the extirpation wasn't due to lack of regulation. In 1215, the Magna Carta called for the king's salmon weirs to be dismantled in order to preserve fish for public use. Richard the Lionheart passed a statute requiring that weirs and other trapping devices avoid spanning rivers and streams—maintaining gaps so that some salmon could reach their spawning grounds. In 1424 Scotland's James I banned weirs and traps entirely and established a period during each run that was closed to fishing.

But throughout Europe, enforcement was nonexistent. If the fish were there, people took them. After all, no one owned them.

Overharvesting was followed by pollution during industrialization, along with major public works projects to "improve" streams: cleaning rivers of the downed wood that creates fish habitat, straightening stream channels, and building levees to prevent floodwaters from wandering around. The combination of overfishing and habitat destruction was effective. It eliminated a key source of inexpensive protein, along with one of the continent's last traces of wildness.

As harvests began to decline in western Europe, the industry turned to Atlantic salmon populations that bred in New England, the Maritime Provinces, and Quebec. Salmon were commercially harvested from rivers that emptied into the northwestern reaches of the Atlantic Ocean, ranging from Long Island Sound in the south to Hudson Bay in the north. These fish matured in the Labrador Sea and Davis Strait off Greenland's west coast, side by side with the populations that bred in Europe. Conservative estimates put annual runs of the North American Atlantic salmon at 5 to 12 million fish annually. The Pilgrims ate salmon frequently; Captain John Smith and Henry Hudson both noted their abundance. In 1672, an angler in Nova Scotia reported that catching 3-foot-long salmon was routine and that 6 footers were possible. Even with a 40-percent fish-story discount, those were large animals.

Historically, fishing has been conducted in what biologists call a scramble competition: if no one owns a resource, whoever gets there first and takes the most wins the competition. Sociologists refer to the result as a tragedy of the commons, based on the history of publicly owned pastures in the American colonies: farmers who overstocked benefited, even though the quality of the pasture declined and depressed the total income realized by the community. The problem was that anyone who showed restraint lost out, even though everyone would have benefited if everyone had shown restraint.

A scramble competition for the Atlantic salmon of North America's eastern seaboard played out over a century. Predictably, it ended in a commons that looks tragic. Runs are gone from many rivers. Commercially, Atlantic salmon are extinct in most of their former range.

The story continued when Europeans arrived in the Pacific Northwest. William Clark of the Corps of Discovery saw Columbia River salmon in 1805 and wrote: "The multitude of this fish is almost inconceivable." Recent estimates suggest that 11 to 16 million fish once bred in that river's watershed alone. An early white settler of south Puget Sound named Ezra Meeker reported that during the 1870s, two men in a boat could regularly catch 1,200 pounds of salmon a night in western Washington's Puyallup River. In the same decade, the fishing industry was cranking out 630,000 cases of canned Columbia River salmon a year, with each case containing forty-eight 1-pound tins. Often, canneries along the Columbia and in Puget Sound couldn't keep up with the catch, and half or more of the harvest would be wasted—the carcasses dumped.

The pressure continued through the 1970s to 1990s, when fishing interests regularly took 60 to 90 percent of the annual runs. By then, the Columbia and its tributaries had been dammed in more than sixty places; the human population had exploded, and streams and estuaries throughout the region were being ruined by helter-skelter development. Salmon returns in Pacific Northwest rivers are now estimated at less than 7 percent of historical levels.

If anything, the situation is even worse in Japan, where masu, chum, and pink salmon were once abundant. An estimated 98 percent of streams throughout Japan have been dammed or otherwise altered. Most of the remaining salmon runs are on the northernmost island of Hokkaido, and many of those populations are on life support, born and bred in hatcheries. In the Pacific, only salmon native to northeast Siberia and some regions of Alaska still have healthy populations.

~

So as we plant trees and restore degraded stretches of Tarboo Creek, we're aware that we're fighting a thousand years of history. Restraints on salmon fishing have never been enforced effectively—except by the Ainu people who had lived on Hokkaido for millennia before the arrival of Japanese settlers, Native American tribes before contact with Europeans, and countries like Iceland.

In Puget Sound today, people get upset about the nearshore fishing—like the boat in Tarboo Bay—because they can see it. But we have only a faint idea of what fishermen are taking out in the open ocean. We can't see them, though they can use sonar to see the fish. There are ongoing and bitter disputes between the United States and Canada over how many of which salmon species can be caught where—just as there were bitter disputes in the 1970s and '80s between those two countries and Denmark, which claims jurisdiction over the Atlantic salmon living off Greenland's coast.

The situation is becoming even more complicated as a result of uncertainty about what changing water temperatures and ocean acidity will do to salmon populations. Historically, salmon numbers in the Northwest have fluctuated in response to what oceanographers call the Pacific Decadal Oscillation—a natural twenty-to-thirty-year-long cycle in ocean temperatures. Things will be good for salmon in the Pacific Northwest for twenty to thirty years, then bad for twenty to thirty years. Another well-established pattern is that when conditions are good for the fish native to northern California, Oregon,

Washington, and British Columbia, things are tough for Alaskan salmon, and vice versa—though I've yet to see a good explanation for the contrast.

The current uncertainty created by climate change hinges on two of a salmon's most basic needs. The first is healthy streams for nesting and rearing. Based on the IPCC's projections for warmer, wetter winters and hotter, drier summers in the Pacific Northwest, researchers are forecasting more winter flooding along with lower water levels and higher water temperatures in summer. Winter floods can excavate redds before young salmon fry hatch; high water temperatures in summer can stress juvenile salmon. The implications for breeding success in the Northwest are not good. But as the Arctic Ocean becomes more free of ice, salmon are beginning to colonize rivers that empty north of the Bering Strait. Salmon, like trees, are moving poleward. No one knows whether gained breeding habitat in the north will balance out lost breeding habitat in the south.

The second basic need is food once salmon reach the ocean. Prey density depends on the abundance of photosynthetic bacteria and algae in the water, which depends on the availability of nutrients. Most of the ocean is a desert, biologically, because it is so nutrient poor. Seawater lacks abundant nitrogen, phosphorus, and iron for a simple reason: when marine creatures die, their nutrient-laden bodies sink to the bottom. There, their remains are slowly passed around the diaphanous food webs that exist in the lightless, pressure-packed, and near-freezing expanses of the ocean floor. Eventually, one of three things happens. The atoms can be carried deep into Earth's crust on a subducting plate, in effect disappearing forever. Alternatively, the carbon and hydrogen in the tiny corpses may gradually turn into petroleum, then be pumped out of the ground and burned in cars. The final option? An upwelling current can bring the elements back to the surface, where they can be used again by organisms.

The ability of water currents to sweep material up from the ocean floor to the surface depends, in large part, on changes in the density of water layers—how heavy water is per unit volume. In general, colder water is heavier than warmer water. So if the surface of the ocean cools, the water will tend to sink and be displaced by slightly warmer, slightly lighter, and much more nutrient-rich water from below. When this happens, production goes up. But if the surface warms, the top of the water column gets lighter and more resistant to upwelling, and thus nutrient poor. Production goes down.

Based on these patterns, climate change will probably be bad news, overall, for salmon and other fish populations. But things are complicated. In some places, winds are increasing due to warming air. Strong and sustained winds can move enough surface water aside to force water up from below, causing upwellings that increase productivity. And as the northern oceans warm, the distribution of copepods—tiny crustaceans that feed the fish that salmon eat—is changing. Warm-water copepod species are moving north, and in some regions overall biodiversity among copepods and other planktonic forms is increasing. It's not yet clear how these changes will affect salmon and their prey.

The orca whales that feed on salmon also bear watching. Puget Sound's orca pods were devastated in the 1960s and '70s by captures for marine shows and aquaria—the tourist trade. Their numbers have rebounded somewhat since but are sensitive to changes in salmon abundance. They will be affected by climate change as well.

An ecosystem is a tapestry; climate change pulls at the threads.

⁓

In the meantime, it's become clear that the classic solution to declines in salmon populations—hatcheries—has failed. Hatcheries are located on or near salmon streams; the staff catch returning adults, strip them of their eggs and sperm, and combine the gametes in buckets. When the fertilized eggs hatch, the fry are kept in predator-proof pens and fed as much as they can eat until they are ready to migrate to the ocean. The idea is for artificial rearing to increase the survivorship of young fish and make up for the widespread loss of breeding habitat.

Hatcheries on the Columbia River released 4.5 million fry a year in the late 1890s and were cranking out as many as 120 million a year a century later. Still, runs crashed. Something was wrong with the promise that artificial rearing on an industrial scale would make up for the loss of spawning grounds to dams, channelization, and suburbanization. The story is the same throughout the Pacific Northwest, where an estimated 5 billion salmon are released from hatcheries each year. A 1991 assessment concluded that 33 percent of all historical populations were extinct and that half of the runs remaining in the region were endangered. The fish are gone from a third of the geographic area occupied in 1850.

Based on more than a hundred years of data like these, the claim that hatcheries can make up for habitat loss is false. Knowing what we know now, hearing someone advocate for hatcheries is like listening to a doctor advise a patient that it's okay to eat junk food, abuse alcohol, sit all day, and chain smoke because we have state-of-the-art emergency rooms and ICUs that can solve any health problems that might result. But the jobs and sport fisheries supported by hatcheries have created a vested interest group that fights any attempt to close them.

Not only have hatcheries failed to solve the problems caused by overfishing and habitat loss, but also in some cases they have actively made things worse for the remaining wild populations. Salmon evolve rapidly in the hatchery environment, resulting in fish that are less well adapted to natural conditions than their recent ancestors were. In British Columbia, for example, Chinook salmon lay smaller eggs, on average, in rivers that are heavily supplemented with hatchery-produced young than do nonsupplemented populations in otherwise similar streams. Small eggs are favored in the hatchery environment, but larger eggs do much better in the wild. So crossbreeding between hatchery and wild stocks is making salmon less capable of living a natural life.

David Montgomery recognized the root of the problem: the success of hatcheries has always been measured in numbers of fry released, not the health of the populations they are supposed to be supplementing. The situation reminds me of farming in Wisconsin when I was growing up: the same newspaper's front page would tout record corn production in bushels per acre but then profile families that were losing their farms due to the high costs of capital-intensive farming and depressed prices due to overproduction. We were using the wrong yardstick to measure success.

Hatcheries are also expensive. An analysis done in Oregon in the early 2000s estimated that each adult fish produced from hatchery-reared young cost at least $14 in years with good survivorship and up to $530 in years with poor production; a more recent analysis of hatcheries in the Columbia River Basin put the average cost at $73 per adult fish. If the money spent on hatcheries over the past 130 years had been spent on habitat protection and restoration instead, the situation would be far different today.

Commercial fishermen and fisheries managers are not stupid, and the industry and its regulators may yet find a way to manage salmon populations sustainably. One proposal is to allow individuals to own a percentage of the total allowable catch. The intent is to take the scramble out of the competition and the tragedy out of the commons. As an alternative or in concert, David Montgomery and others propose that salmon fishing be allowed only at river mouths, where the health of the runs can be monitored precisely. This is how the people of North America's First Nations traditionally managed salmon populations; their system worked for more than thirteen thousand years. Iceland also manages its salmon runs on a river-by-river basis.

Something has to change, or the story of salmon in the Pacific Northwest will play out the same way it did in Europe, New England, the Maritime Provinces, and Japan. The old ways have failed.

Planting Season

~~~~~~~~~~~~~~~~~~~~~~~~~

It's a great day when the trees arrive at the start of planting season. As you unload bundles of freshly dug bare-root plants, you never think of failure—of yellowed cedar saplings fried by an early spring heat wave or needleless Doug-fir killed by mice that gnaw a ring around the stem or saplings that simply fail and die for no apparent reason. Instead, you breathe in the delicious evergreen scent and imagine each tiny sapling as a towering spire, 250 years old and 4 feet in diameter.

After planting about three thousand one-or-two-year-old trees and shrubs in the winter and early spring of 2005 with help from our plant-a-thon friends, we've been adding about five hundred more each year. Many of the plants are bare-root stock purchased from a nursery run by the nonprofit Washington Association of Conservation Districts, but on occasion we grow our own or dig little firs and cedars from our land for transplanting.

We sometimes get our trees with NWI's annual order from the big nurseries, our hundreds to their thousands. NWI's field crews and plant-a-thon parties have replanted more than 240 acres in the floodplain just downstream from us on a long-defunct dairy farm. The group purchased the land using grants from a U.S. Fish and Wildlife Service program that funnels money from excise taxes on fishing equipment and motorboat fuel into preservation of coastal wetlands. It's a way for the beneficiaries of healthy fish populations to help make sure those populations stay healthy.

When we pick up bare-root saplings from the nurseries or dig our own little trees, the first step is to store the plants until they can be planted out, by covering the bare roots with soil or sawdust or wood chips. When they are heeled in in this way, lines of rust-and-black vine maple and red elderberry alternate with emerald sprays of western redcedar and Douglas-fir. The plants come from the supplier wired into bundles of twenty-five or fifty, so the deciduous shrubs poke up from the ground like bristle brushes; the conifers form fragrant bright-green mounds.

In many years, we also harvest alder pull-ups—pencil-thin red alder trees, 2 to 3 feet high. When the soil that little alders are growing in has been thoroughly wetted by winter rains, you can pull the entire plant, roots and all, right out of the ground. If you can find an old logging road where alders have seeded in recently, you can collect bucketsful in minutes.

Red alders are a pioneering species—meaning they can take the high temperatures, dryness, and poor soil quality that prevails on sites opened by windstorms, fires, logging, or construction. The reason they can cope with poor soil quality is that their roots are decorated with little orange baubles that fix nitrogen. Inside each of these nodules are millions of symbiotic bacteria from the species *Frankia alni* that convert molecular nitrogen from the atmosphere—dinitrogen or N_2—into what a biochemist would call reduced or fixed forms of nitrogen, like nitrate (NO_3), nitrite (NO_2), or amino groups with the chemical formula $-NH_2$. Although molecular nitrogen is superabundant (N_2 makes up most of the air we breathe), chemically it's so stable that it's almost inert. Nitrogen atoms can only participate in key chemical reactions—like the ones that build the proteins, RNAs, and DNA that make life possible—after they have been fixed. The fixation process is impressive biochemistry: it depends on breaking the strong, rigid triple bonds that glue atmospheric nitrogen (N≡N) together and substituting single bonds to hydrogen or carbon atoms, like H–N–H. Human chemists couldn't come up with an efficient way to do this on an industrial scale until the 1910s; the scientists who did the key work won Nobel prizes.

Frankia and other nitrogen-fixing bacteria do this chemistry with a sophisticated complex of enzymes collectively called nitrogenase. Keeping this machinery humming is an expensive, energy-demanding process for the bacterium and is rarely possible without help from a photosynthetic host providing the requisite power supply. So in many cases, a plant partner like red alder pipelines a steady stream of energy-rich sugars in exchange for little jewels of reduced nitrogen from *Frankia*. The arrangement means that alders have a ready supply of fertilizer manufactured in their roots. It's a business partnership that has endured for millions of years.

~

We're trying to reforest land that was wooded for millennia but has been cleared of trees for at least 120 years. A photo of the Tarboo watershed taken in the 1890s shows the aftermath of the initial cut: people scratching out a living on homesteads surrounded by muddy fields dotted with the charred snags of old-growth trees, with intact forest receding in the background. The scene reminds me of roads I've walked in Ecuador and southern Mexico where the initial deforestation is just being completed. There, destitute families of squatters live in shacks, trying to grow maize on steep slopes or raise a few cattle in pastures where shade is provided by one or two big trees—relics from the rainforest.

In Puget Sound, the early white settlers were focused on farming. They felled trees in lowlands, where the ground was relatively flat, and burned the downed trunks to clear fields. There was some commercial logging, but it was concentrated close to the coast. In those days the timber had to be pulled by oxen or horses to big rivers and floated down to the sea, or pulled to the ocean itself, where the logs were assembled into booms and towed to the nearest mill. The process of pulling downed logs from a forest to a yarding area, where loads are organized into decks for longer-distance transport, is called skidding. The steep sections of the street called Yesler Way in downtown Seattle were a skid road for a waterfront mill in the city's early days; the prostitutes and penniless men who lived along it helped inspire the term *skid row* for slums throughout the United States.

A later photo, from the 1930s, shows our place and surrounding valley still treeless. By then virtually all of the Puget Sound lowlands had been cleared by commercial logging operations. The companies had long since abandoned animal power; instead they moved the downed logs with cables attached to huge steam-powered engines called donkeys. The donkeys pulled the logs to yards where they could be loaded onto railcars and taken to the mill.

That 1930s-vintage photo of our place also shows a cedar shake mill in operation. Western redcedar is the preferred wood for the shakes and shingles used in roofing, as well as clapboards for siding and material for outdoor decks, because it's so rot resistant. But only old-growth cedar has no knots and thus can be used for shakes and shingles. The location of the mill seems odd, then, because there are no big trees in the picture. It's possible that the raw material for the mill came from forested land miles away and upslope; by then the big companies were cutting up the sides of the Olympic and Cascade mountains and trucking logs down. But there may have been local sources as well: cedar could have been scavenged from stumps and the odd downed log left from the original cut or harvested during the early days of tractor logging. When the first tractor crawlers were available, independent operators called gyppos used them to reach old-growth trees in inaccessible areas that had escaped the initial cuts in the late 1800s. The gyppos would run the tractors up streambeds, cut the big trees that remained, and then skid the logs back down the creek. Stream channels were also the preferred skid road in the horse-and-ox days. Both eras were dark days for salmon streams—the animals and logs and tractor treads would tear up the channels, and then winter rains would erode the bare hillsides and fill gravel beds with sediment.

～

The deforestation of the Tarboo Valley from 1880 until 1930 is a speck on a map—a tiny data point in a trend that began three thousand years ago and continues today. In the Bible, Psalm 92 says that the righteous shall grow like a cedar; Isaiah 41 and Ezekiel 17 maintain that God planted cedar in the wilderness and the mountains. None of the verses kept the trees from being cut—King Solomon built his temple from the cedars of Lebanon. Throughout the Mediterranean, tree felling and grazing by goats have kept landscapes treeless for millennia; the cedar of Lebanon adorns that country's flag but is endangered there. By the Middle Ages the axmen had moved north: cutting to convert land to agriculture and for firewood and charcoal production was clearing Europe of trees. Old-growth forests were essentially gone from that continent by the late 1600s.

After Columbus and the *Mayflower*, the wave of deforestation jumped the Atlantic and swept across North America. Settlers led; large logging companies followed close behind. Between them, virtually every large stem in the United States was cut in less than three hundred years. As the Englishman Michael Williams has documented in his seminal book *Americans and Their Forests*, the process started in New England, progressed to the Lake States, moved to the

Deep South, and then jumped the Great Plains to the Rockies and the Pacific Northwest. Almost 40 million acres of American forests were converted to farmland between 1850 and 1859; from 1860 to 1880, 5 million acres were cleared in the state of Wisconsin alone. Over a quarter of Ohio's counties were 80 percent cleared of forests by 1880; ten years later, half of all the counties in that state were essentially deforested. The "inexhaustible" pineries of the Lake States—Michigan, Wisconsin, and Minnesota—were gone, for all intents and purposes, by 1900. Thanks to inefficient rafting and saws, the practice of cutting large trees 4 to 6 feet from the ground, and forest fires that burned out of control, much of the wood was wasted. In Wisconsin, so many branches and other types of slash were left on the ground after logging that when it dried and fires swept through, the heat was intense enough to vaporize the organic material and kill the microorganisms in the soil—effectively sterilizing it. The cutover land still hasn't recovered.

Almost no one spoke out against the waste or the loss of irreplaceable older forests; resources were there to be exploited. The concept of sustainability hadn't been invented; ideas like the land ethic were unimaginable; no one thought their grandchildren might want to see what the big woods looked and felt like. Everything got cut. After the move to the Rockies and the Pacific Northwest, the only thing that stopped the wave of deforestation was the Pacific Ocean.

All of this cutting was done with handsaws and axes; the portable chainsaw wasn't invented until the 1950s. With that tool, one person can do in an hour what used to take two men two days. And now the chainsaws have moved south, to the tropical rainforests.

～

Today the United Nations Food and Agriculture Organization (FAO) is responsible for tracking the fate of the world's forests. It does this by analyzing satellite images. During the 1990s the FAO estimated that net forest losses averaged about 4.1 million hectares per year, worldwide. (A hectare is about 2.5 acres or roughly the size of two American football fields.) To put this number into perspective, 4.1 million hectares is the equivalent of two Massachusetts and represents 0.11 percent of the world's total forestland. In a single decade, then, about 1 percent of the forest cover that existed worldwide vanished.

This isn't the entire story, however. Statistics on net losses don't change when ancient forests are cleared and replaced with scrubby second growth or with commercial oil palm or teak plantations. In many cases we're swapping old, diverse forest landscapes for new monocultures and exchanging native trees for species that are exotic to the area. The net figures also hide a north-south divide: extensive loss of species-rich tropical forest is masked by an increase in younger, much-less-diverse forest types in northern latitudes. We lost an average of 6.3 million hectares of tropical forest per year during the 1990s, but the global net loss of forest was much less than this because of regrowth in the temperate, species-poor woodlands of China, Europe, and the United States.

Another issue with taking the numbers at face value is the way the FAO defines forest: as land with 10 percent or more canopy cover by trees. Most people wouldn't look at a landscape with 10 percent tree cover—or even 20 percent or 30 percent—and consider it forest; neither would most other species. If the FAO defined forest as an area with 80 to 100 percent tree cover, the number of hectares that would be considered forest would be much smaller and the percentage-of-forest-loss estimates would be much larger. The definition also hides the impact of selective logging, usually sponsored by large companies that remove mahogany and other high-value trees for export to Europe or the United States. The roads built in these operations open pristine woodlands to poachers and squatters, and increase the risk of forest fire.

Finally, the FAO's figures don't include the effects of forest fragmentation—of carving large, contiguous blocks of forest habitat into small bits surrounded by farmland or suburbs. In Brazil, researchers did this experimentally and documented two striking changes over time. First, large animals that demand a lot of space—jaguars, tapirs, and army ants—disappeared from the fragments almost immediately. Second, habitat quality declined dramatically within five years. This happened because so much edge was exposed to the surrounding pastures and croplands. The exposed forest edges dried out, large trees died and fell, and vines and weedy species from the surrounding developments took their place. The researchers found that fragmentation affects 1.5 times the area actually cleared.

We see the same thing happening at the edges of clear-cuts in the Tarboo Valley: the trees that form the border of the remaining intact forest have never been exposed to intense sun and wind. Many can't change their needles and root systems fast enough to cope with life on the edge, and they die.

～

According to the FAO's latest report in 2011, the overall situation worsened from 2000 to 2005—the most recent interval with data available. Globally, net losses of forest increased during that period to 6.4 million hectares per year, up from the 4.1 million hectares per year in the 1990s. Most of this increase occurred in the tropics, where net losses jumped from an average of 6.3 to 8.0 million hectares per year. The total area of tropical forest lost each year in the early 2000s was larger than the country of Panama; the daily loss rate was 219 square kilometers. On a two-lane highway, it would take you more than forty minutes to drive the perimeter of the area cleared each day. The rate of tropical forest destruction is accelerating as large companies convert forests with two goals in mind: establishing oil palm plantations to meet surging global demand for biofuels, and planting soybeans for animal feed needed by beef producers trying to satisfy increasing demand for meat worldwide.

The longer-term outlook is not good. More than 13 percent of the Amazonian forests are already gone, and less than half of the forests that once covered southeast Asia remain. Deforestation rates are increasing in Indonesia and Malaysia and in central Africa. If present trends continue, most of the world's tropical forests will be wiped out by the year 2100.

Our older son worked in the Brazilian Amazon just long enough to become deeply discouraged about its future. For much of the season he was there, it was smoggy from fires set by large landowners who wanted to convert forest to pasture for beef cattle or to fields for soybeans and sugarcane. Fire-prone seasons in the American tropics usually occur when warm ocean waters move close to either the Pacific coast or the Atlantic coast and keep humid ocean air from moving over land, lowering rainfall. As the oceans warm in response to climate change, droughts in the Amazon are projected to become more frequent and severe, encouraging fires to burn out of control. Recent maps show Brazil's "arc of deforestation"—a boomerang-shaped region with intensifying fire and industrial-scale cutting. Even though the vast majority of the logging is illegal, the arc is expanding to the north and west.

In the United States, the year 2000 was among the worst on record in terms of losses to forest fires: almost 3.4 million hectares burned. But in 1997 to 1998, 3 million hectares burned in Bolivia alone. Indonesia lost 8 million hectares that same year; in just one of Brazil's Amazonian states, 5 million hectares went up in flames. The total loss of tropical forests that year was equivalent to half the state of California.

Recent work has shown that forest fire frequency is increasing outside of the tropics as well—creating what scientists call a positive feedback effect on climate change. The term is tricky—*positive* in this context doesn't mean good; it means the feedback adds to or speeds up the process. When forests burn, CO_2 is released and the total atmospheric concentration of CO_2 rises. Recall that CO_2 molecules, in turn, are efficient at absorbing infrared radiation that has been reflected from Earth's surface, so that instead of being lost to space, the energy stays in the atmosphere and warms the planet. Forest fires increase CO_2 concentrations in the atmosphere, which increase the probability of warm ocean currents and drought, which increase the chance of forest fires, which increase CO_2 levels in the atmosphere, and so on. Fires create a positive feedback loop.

But there are also negative feedback effects on climate change. In this context, *negative* doesn't mean bad; it means the feedback reduces or slows down the process. For example, plants use CO_2 in the atmosphere as a raw material in photosynthesis. The carbon in carbon dioxide is used to make sugars and eventually cellulose, lignin, proteins, and nucleic acids—all the stuff that makes up leaves, roots, tree trunks, and strawberries. When CO_2 increases in the atmosphere, then, it should act as a fertilizer. This prediction can be tested experimentally. If researchers augment CO_2 concentrations

by pumping it into the air above forests or prairies, productivity usually increases—plants get bigger. More sugars and cellulose and other compounds in plants means there is less CO_2 in the atmosphere, so increased plant growth reduces the impact of greenhouse gases. It's a negative feedback—it turns the thermostat down.

The interactions between positive and negative feedback effects are part of what makes predicting the extent and nature of climate change so complex. But the IPCC reports have been consistent, and consistently correct: overall, the climate is warming. Forest fires will be more of an issue in the future, not less.

It would be cynical, though, for Europeans or Americans to lecture biologists and policymakers from the tropics about the evils of deforestation. Brazil's arc of deforestation is analogous to the Mediterranean-to-Baltic cutting that occurred from biblical times to 1700 and the Atlantic-to-Pacific wave of forest loss that swept across North America from 1700 to 1990. What happened in the Tarboo Valley in 1895 is happening in Rondônia, Brazil, and Irian Jaya, Indonesia, today. Scientists from the industrialized countries are in the position of saying, "Do as we say, not as we did."

On the other hand, research is showing that forest loss is probably not in the best interests of the tropical nations themselves. For example, recent work by European and Brazilian researchers has shown that incomes increase in the early years of forest conversion but then decline once the forest is gone. There is a boom, then a bust. The problem is that the productivity of cattle pastures and cropland declines in a matter of a decade or less as the thin tropical soils become exhausted. The bust leaves deforested areas in the same economic state as before—but without the timber, food, and water provided by the forests. An array of economists and biologists is now advocating agroforestry systems and sustainable logging based on native species as the best way to increase incomes long term. If further research supports the conclusion that sustainability pays, tropical nations may not need more than three hundred years, as the "advanced" countries did, to reach the conclusion that forests are good.

~

My forester friend Mike Cronin likes to reminisce about a logger named Jim Johnson, whose son still lives in the Tarboo Valley. Jim made his living cutting trees. But in the 1930s, long before almost anyone else in the Pacific Northwest had the idea, he also replanted them. At the same time but half a continent away, Carl Leopold was planting pines with his family in Wisconsin. Before he retired, Jim harvested the Douglas-fir he'd planted as young man; in his mid-sixties, Carl built a home from trees he'd planted as a teenager. Both men were pioneers.

Carl planted trees throughout his life. In the 1950s and '60s he planted pines with Susan near their home in Indiana, and in 1993 he initiated one of the first efforts to restore native rainforest vegetation in the tropics—on 145 hectares of derelict cattle pastures near the Pacific coast of southwest Costa Rica. Within five years, the fast-growing, weedy tree species he'd planted in Costa Rica were adding the height of a basketball hoop each year; some had trunks the diameter of a field-goal post. He'd planted them as a cover crop to provide shade and build soil for slower-growing, longer-lived trees planted underneath. After ten years, the canopy had closed; seeds deposited in monkey, bat, and bird droppings and carried in by ants had introduced almost a hundred native understory plants. Carl and his wife, Lynn, monitored the progress with instruments that allowed them to measure tree height, diameter, shape, and biomass. He authored some of the first studies ever published on the growth rates of native tropical trees and changes in species diversity in restored rainforest.

There are other reforestation efforts in the tropics—many others. Since 1977, members of the Green Belt Movement founded by Wangari Maathai have planted more than 45 million trees in Kenya. The movement employs local women to establish and maintain tree nurseries; movement staff also run seminars on civic engagement, women's rights and reproductive health, and sustainable economic development. The success brought Maathai the 2004 Nobel Peace Prize and inspired the United Nations Environment Programme (UNEP) to start the Billion Tree Campaign with the aim of supporting a billion new trees planted worldwide each year. Fewer than five years after its inception in 2006, more than 12 billion trees had been planted and registered on the campaign's website.

In Borneo, Willie Smits leads a group that is restoring rainforests at a 2,000-hectare site called Samboja Lestari. After being cleared by large logging companies and devastated by repeated fires, the area had become the poorest community in the district. In 2002, Smits's organization, the Borneo Orangutan Survival Foundation, began planting 740 different species of trees in a design meant to discourage wildfire. Fire-resistant sugar palms—with sap that can be tapped to produce sugar and ethanol for fuel—were planted in large rings around diverse plantings of native trees. By 2006, the replanted sites were providing habitat for orangutans and yielding crops, firewood, and timber for local people.

Similar projects can be found on almost every continent. Deforestation has led to a full 38 percent of China's total land area being classified as badly eroded; in response, the government instituted massive tree planting programs—though in some or many cases, exotic trees are being planted in monocultures. Even the cedars of Lebanon are making a comeback. Although replanting programs in Lebanon have been suspended due to political instability, Turkey has replanted more than 65,000 hectares in cedars since the mid-1980s. Oils distilled from the wood and seeds of these trees were used by the ancient Egyptians as a preservative in mummification; recent research suggests that the compounds may also have potential as a safe and biodegradable insecticide.

There are thousands of other projects, glowing like candles in the night. A scientific field called restoration ecology is flourishing, producing data on the best ways to replant forests and grasslands, reintroduce animals, and restore high-functioning ecosystems. We've come a long way since the first attempts at reforestation and ecological restoration, initiated by people like Jim Johnson, Aldo Leopold, and Carl Leopold.

So we are at a crossroads. A three-thousand-year-long wave of deforestation will probably be completed this century. And compared to what a well-capitalized investment group can do to a tropical rainforest in the span of a few months, reforestation efforts like ours seem like throwing pebbles into the ocean. But when thousands of people are throwing pebbles in thousands of places at the same time, things change. The rings from the splashes are expanding, reclaiming lost landscapes. Planting a tree is a way to apply hope. In restoration is the preservation of the world.

⌢

The first thing to understand about reforesting an abandoned pasture, like the 6 acres east of the creek at our place, is that it's hard, even in areas that don't have clay soils like our Dead Zone. One of our boys found a research report that mentioned an old pasture in Washington's Olympic National Park, surrounded by mature forest, that had not been grazed or hayed for eighty years. The field still had no saplings growing in it—even though tens of thousands of tree seeds had been raining onto it for the better part of a century.

Grass is tough. For small trees to have a prayer of surviving in it, you have to do something to break the sod. If you can scavenge old sheets of cardboard and lay them on a planting site, the grass will be smothered a year later and ready for a tree. Herbicides can be effective but have to be applied carefully; one year a crew from the county got too close to Tarboo Creek with their sprayer and treated both sides of the bank for a long stretch; the denuded sides gave way in the following winter's floods and sent truckloads of sediment careening toward Tarboo Bay.

The Leopold family used to hire a neighbor to come in and plow, turning over the sod to discourage grass growth in that year's planting sites. Many times, we mimic this approach by scalping a small rectangle of sod. We do this by taking a stout, broad-bladed hoe—an industrial-strength tool used for clearing fire lines—and whacking at the ground when it's good and wet. Two or three thumps like this and we appreciate why tree seedlings have such a hard time. The grass roots form a netting so thick and tough it's like peeling back a 3-inch-thick shag carpet. The sod sucks up every possible water molecule and every available nitrogen, phosphorus, and potassium atom, leaving nothing for little tree seedlings to grow on.

Grasses are particularly effective competitors because when soils dry out in late summer, the blades simply go dormant. They start growing again when the rains return in late fall, or they just wait for spring. But young trees can't do this. They still have green leaves exposed, and the leaves have to perform photosynthesis so the plant has something to eat. But to do that, they have to open pores in their surfaces and take in carbon dioxide. When the pores are open, water escapes. If it can be replaced from soil water taken up by the roots, the evaporative loss from leaves is actually a good thing—it cools them down and prevents overheating. In effect, plants sweat. If they can't, and if temperatures spike, they can die of heat stroke.

So come early September, when the Northwest's annual ten-week drought is starting to bite, the grasses have used up all the water near the soil surface and gone to sleep. Big trees can cope—they have roots deep below the surface and can easily find enough water to keep their leaves cool and moist. But saplings, with roots that barely penetrate the sod layer, are in a bind. All it takes is one good stretch of hot weather and they can burn up.

This is why it's much easier to reforest a clear-cut than a pasture. Most saplings planted in a clear-cut live, and most will be two to three times bigger than seedlings planted in a meadow at the same time—if the little trees in the grassland survive at all.

~

Much of our planting has been focused on the old pasture just east of the creek and in the floodplain beside the newly meandered stream. But we've also been planting in the wooded uplands at our place, where all the big western redcedars and firs were logged off. There we're replanting among broad cedar stumps, in the shade of tall bigleaf maples that were spared. The maples were damaged in the recent logging operation, however—by soil compaction, blows from falling cedars, and collisions with the skidders and other logging machinery. So they are not thriving. Large branches in their crowns are dying, rotting in place, and then snapping off in winter winds. The branch ends often plunge straight down and plant themselves a foot or more deep in the ground—as if they grew there. We've never seen this happen—we've only observed the aftermath—but I imagine the branches humming like tuning forks when their ends strike the ground.

When we underplant in the woods, there is no grass to scalp. Instead, the challenge is to find a root-free zone that can take a sapling. Stick a shovel in the ground in the Northwest woods and you'll appreciate why it's rare to see a young tree growing in the soil. Just under the surface of the forest floor is a mat of fine, tightly woven tree roots. The mat is in the black-soil zone, where roots are recycling nutrients released from recently fallen leaves and branches. The depth and density of this surface-root layer rivals what you find in a grassy pasture. If you hack through this mat and keep digging, you find stout root cords crisscrossing the deeper soil layers, branching into any pocket of soil that holds water and nutrients. In comparison, a sapling's roots are slender and tender. In an arm wrestle over water, they don't stand a chance.

The saplings find a way, though. In old forests, seeds germinate and start growing on nurse logs—downed, rotting trunks with upper surfaces 3 feet or more above the root mat. Once an old tree has been down for a decade or two, it's common to see a row of 4-foot-tall hemlock or cedar saplings lined up along its back, like schoolchildren waiting for the bell.

The trees that emerge from the row and shade the others out will send their roots careening down the length of the nurse log, following the rotting cambium layer that held the old tree's vascular tissue. This is where the lion's share of the nutrients lie in the log. But some will head down, strike the soil, and take root there as well, straddling the log like a cowboy in the saddle.

The same thing happened when the early loggers left 6-to 8-foot-tall stumps. Seeds germinated on the flat surface of the cut, and the saplings that survived sent roots down the sides of the flaring column. When the carcass rotted away underneath, the great tree that remained would be perched on a web of roots—standing on tiptoe.

So when we carry a bucket of seedlings through the uplands, we have to look for open patches. Spaces a few feet from stumps are good, because the roots from trees felled five years before are rotting and roots from nearby trees haven't yet taken their place. If you look up and see sky, you have checked both boxes: there is space for the roots and space for the stem. Then you can dig.

～

Planting can bring pleasure or pain, depending on the soil you're working in and the tool you're working with. A planting tool should be chosen carefully. You want a shovel with a long, narrow blade. Depending on your preference and the soil involved, you may even want a hoedad: an implement with a stout wooden handle and a tonguelike steel blade making an L. If you choose a shovel, you want it to have a long throat—a metal tube that hugs the wooden handle. You never mean to pry, but sometimes you can't help yourself when you're planting. The handle of a tool with a long throat is less likely to snap if you get careless and use the shovel like a lever.

Saplings that are intended for replanting clear-cuts are grown and sold as plugs. Plugs have compact root systems and can be planted with a hoedad without digging a hole. Crews of professional tree planters work through clear-cuts carrying hundreds of saplings in a cloth bag—to keep the roots moist—slung over one shoulder. They plant a Douglas-fir plug every five paces or so by swinging their hoedad's blade into the ground, prying open a vertical slit in the soil, pushing the plug in, stomping the opening closed, and moving ahead another five paces. The sequence takes ten to fifteen seconds; a good hand can plant three or four thousand trees in a day. Some professionals have planted a million trees this way.

Saplings intended for underplanting in existing woods, or for reforesting pastures or hayfields, are a different animal entirely. They are field grown for two years, so they can be 1 or 2 feet high aboveground with roots a foot or more long. They compete against grass much better than plugs or younger bare-root stock, but the

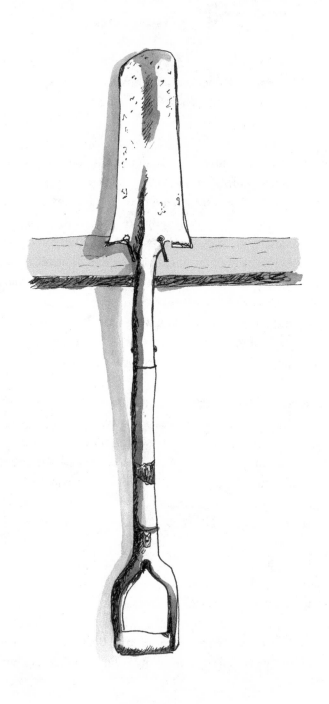

downside is that they are labor intensive to plant—they take a lot of digging. We're usually planting big bare-root stock in the pasture or cutover uplands at Tarboo Creek, so we do a lot of digging. And we learn a great deal in the process.

Some holes turn out to be archaeological sites. I've found shards of glass and crockery, pieces of brick or sheet metal, and cogs or rings from old logging or farming machinery. Every hole has a geologic record, too. It's actually rare for us to see a classic black-then-brown-then-tan layering of soil horizons, in part because of the extent of human disturbance and in part because we have so many clay or gravel pockets—gifts from the glaciers that scoured Puget Sound. When the grade school students planted at our place with their families, we'd find necklacelike rings of cobbles at the base of saplings that had been planted in particularly gravelly holes—shrines to the pain their shovels had just endured.

In winter, the soil in the Dead Zone is saturated with water and sticky enough to put on a potting wheel and throw. Just 6 to 8 inches under the surface, the gray clay is often flecked and streaked with reddish orange. The color is from iron, deposited by bacteria that live in waterlogged, oxygen-free soil. But in the same spot in late August, the clay will have dried out and been fired by the summer sun—baked hard enough to dull the blade of a shovel or even throw sparks when struck by a hoedad.

We also have muck soils—blackened dirt that's rich in waterlogged organic matter. This is a favorite spot for western redcedar trees. They grow pancaked roots in sites like this, as the occasional windblown stem, or tip-up, will attest. Tree roots flatten out and stay near the surface if the deeper layers of the soil are waterlogged year-round. The problem is that saturated soils become oxygen free unless they're exposed to flowing water or some other aerating process. Crumbly soils with good structure, or tilth, have air pockets that maintain an oxygen supply for plant roots. But in muck and peat and clay soils, water fills those pockets and prevents air from penetrating. Once the bacteria and fungi that decompose dead leaves and roots have used up all the available oxygen in these soils, there is no mechanism to replace it. Roots can't grow in these oxygen-free zones. They suffocate. So instead they spread into a pancake at the surface, where they can breathe.

A planting hole can also present small mysteries. For example, I've struck pockets of rich organic matter beneath several layers of brown sands and gravels. Where did the black soil come from? Is it the decayed roots of gigantic old-growth trees, or surface soil that was buried recently by a bulldozer or plow? I've also unearthed little patches of bright rust-red soil, far uphill from the orange-flecked clays of the Dead Zone. For all I know, I could be excavating either the remains of a rusted bucket or an iron-laden rock that was transported from the far north and dumped by the glacier. Some holes reveal buried chunks of charcoal, possibly from one of the settlement-era fires or possibly from a forest that burned a thousand years ago.

~

Perhaps the most important aspect of the soil we plant in, though, is invisible to us. All of the trees in a Pacific Northwest forest—indeed, virtually all of the world's trees—live in association with fungi. Most of us know fungi only as baker's and brewer's yeast, the mushrooms on pizza, the mold on bread, and the cause of athlete's foot. But there are millions of different species. Many are decomposers that make a living from the dead bodies of plants and animals, recycling nutrients that would otherwise be tied up in carcasses. The most important of these decomposers are members of the mushroom-making lineage. Together with a handful of bacteria, the mushroom-forming fungi are the only organisms in the world that can digest lignin—the molecule that gives tree trunks much of their strength. Without these decomposers, the world would be carpeted with dead wood.

But fungi are critical to plants in life as well as death. Healthy trees are coated with dozens of different fungal species, from the tips of their branches to the bottoms of their deepest roots. The situation parallels the human body, which hosts dozens or hundreds of species of bacteria. Our skin is coated with a layer of them, and they fill our guts. We are walking, talking habitats for these bacteria—much like trees with their resident fungi. Most of the bacteria on human skin are harmless hitchhikers; we're merely a convenient place for them to live. But the bacteria in the human gut, like the fungi on plant roots, are crucial to nutrition. Bacteria in the small intestine synthesize vitamins we can't live without and digest food molecules our own cells aren't able to process; bacterial cells in the large intestine also help with water balance. Much of our feces consists of the dead bodies of these symbiotic cells. Change your diet—say, from one that emphasizes grains to one dominated by sugars and oils—and the composition of your gut flora will change in concert.

The diversity of fungi that live on tree leaves and stems was discovered recently, and research has yet to clarify whether most of these aboveground symbionts help their host plant, hurt it, or have no effect. Belowground, the picture is much clearer: most trees and other plants could not thrive without the fungi that live in and around their roots. In some cases, the fungi actually penetrate the interior of the roots, meaning that the partner organism is anchored inside the plant's body. Whether one end coats the root exterior or enters the interior, the body of the fungus fans out from the point of contact with the root into the soil, creating a net that captures nutrients. Plants leak sugars and other photosynthetic products to the fungi at the point where the two partners touch; in exchange, the fungi seek out and transport nitrogen and phosphorus atoms from the soil to the plant. It's an underground economy that benefits both parties.

The key to this system is that the branching network of thin filaments that make up a fungus's body presents an enormous amount of surface area to the soil—much larger than the total surface area available in a plant's roots. The exterior of each fungal filament is studded with proteins that secrete digestive enzymes; the enzymes break down rotting plant material in the soil, creating a slurry. Each filament also contains proteins that are specialized for importing the nutrients released by digesting dead plant tissues, so nitrogen, phosphorus, and other key substances can be transported into the fungus's weblike body. When it comes to reclaiming and recycling nutrients, fungi are the most efficient organisms on Earth.

Trees barter for these atoms—especially the nitrogen and phosphorus. A significant fraction of the sugars manufactured in tree leaves is destined for export to the fungi that wrap the tree's roots. The exchange is reminiscent of merchants who pull sacks of wheat and rice off the backs of camels, haul them into the dim interior of a bazaar, and trade them for a handful of diamonds. Even red alders and other species that house nitrogen-fixing bacteria in root nodules will take up with soil-dwelling fungi. Deny tree roots their fungal partners, and the entire plant will be stunted.

Fungal hyphae can even connect the roots of adjacent trees and facilitate a flow of sugars and other molecules between them. A tree that is in the sun giveth; a tree that is in the shade taketh away. How the sunlit, producer trees avoid donating sugars to nearby shaded trees indefinitely—in effect, getting parasitized—is a question biologists are still researching.

When we're out planting trees, though, we assume that the roots will find the fungi they need somewhere in the soil. We don't add them. A friend who is Washington's leading expert on fungi tells me it's safe to assume that spores from the appropriate root-living fungi are just about everywhere. But if I ever walk by a sapling that's still struggling after several growing seasons, I sometimes wonder if its roots are naked instead of clothed in a sheath of fungi, and the plant malnourished as a result. Sometimes I even deposit a handful of forest soil at the base of small trees like these, hoping I'm introducing spores from the missing root-associated fungal species. It's like eating yogurt or other probiotics if your gut gets upset and stays upset during a long course of antibiotics.

~

Once a hole is dug deep enough and wide enough, success in tree planting hinges on the sapling's roots. For the rest of its life, these will be hidden from us. Years later we may walk by and recall that the roots have become coated with fungal hyphae, or we might be dimly aware that the roots have become grafted to the roots of neighboring individuals, with sugars and nutrients passing between them like gossip. But the world of the soil, like the world of the deep sea, is mostly a mystery. We are out of our depth.

Still, at planting time, roots are foremost, and it's critical to keep them moist. The best planting weather occurs on classic Northwest winter days, when the temperature is 40°F and a gray mist is falling—the kind of weather that can settle in for weeks on end and send recent immigrants back to southern California, vowing never to return. On days like these, the humidity is 100 percent and the roots of planting stock are well satisfied. Even if it's raining, though, we're careful to keep the roots snug in a canvas or brown-paper planting bag, or surrounded by soil in a bucket. On dry days or in a wind, this precaution is urgent.

As a tree comes out of the bag or bucket, it's important to look at the length of the roots and the depth of the hole, and judge whether the roots will fit. If the hole isn't deep enough, the roots will lie on their side and the tree will be J-rooted—meaning the ends of the roots are headed to the side instead of down. If the roots continue to grow like this, the tree will be tippy. Even if the root tips succeed in turning down and spreading out, the tree has had to do extra work at the most vulnerable interval in its life. If you were transplanted to a new job in a foreign culture and were struggling with the food and language and housing situation, you'd rather not break your leg as well. If the soil really is impossible to dig deep enough, I've gotten a jackknife out as a last-ditch measure and trimmed the roots.

So the goal is to give the saplings every possible advantage. Little things count. Plant the tree so that it's vertical and not leaning to one side—that way the stem won't have to waste time and energy righting itself. Depth is another Goldilocks problem: don't plant the stem so deep that part of it is buried or so shallow that the tops of the roots are exposed. And if the depth is off a little, things will probably work out. Plants are tough.

If possible, it's good to spread the roots out in the hole so that they are separated and have a head start on finding soil with a fresh supply of oxygen, nutrients, and water. Then hold the top of the plant and start pulling or kicking soil into the hole to surround the roots. When the first third of the hole is filled, gently tamp the soil with your fingers or boot tip to make sure any large air pockets are filled—airy cavities can dry nearby roots and kill them. Finish filling the hole and press a boot around the base of the stem to pack the rest of the soil and remove any other air pockets. Hold the tip of the stem and pull on it gently as you do this—providing enough tension to prevent the roots from compressing along with the soil.

And with that, a tree is planted.

Before moving on, though, it can be important to protect the young thing. If the planting site is mousey, most planters will slip a plastic tube over each tree and secure the sleeve with a thin stake. The plant protector will prevent mice from nibbling the bark when they're hungry in winter. Tall plant protectors will also help with rabbits, who like to nip off tender shoot tips.

Susan's cousin, who is reforesting an old pasture in the Baraboo Hills of central Wisconsin, has to protect her young oaks, hickories, and white pines from white-tailed deer, which are epidemic. She uses a technique favored by Swedish foresters: spraying the saplings with pig's blood. She buys the stuff dry, mixes it up in water, and hoses the little trees down. It gives the saplings an odd reddish hue but puts the deer off their browse.

In the pasture we're replanting, we also have to think about protecting the trees from grass that will inevitably encroach on the little patches that we've scalped and threaten to overwhelm the little saplings. Each summer NWI sends crews in to cut the grass around plant-a-thon saplings with gas-powered string trimmers. We tend to favor mulching the trees with wood chips or weed barrier cloth, or even with burlap—using landscape staples to pin an old coffee bag on either side of a sapling. We get free coffee bags by the pickup load from a roasting plant near Seattle. But when we can't get a mulch down, we have to let the trees fend for themselves.

⁓

Once a tree is planted and protected, you move on to the next. Planting in a clear-cut is all about rhythm—you'll see crews moving as efficiently as dancers as they plant plug after plug, sometimes glancing up to monitor each other's progress. But planting in grass is labor intensive. A plug can be in the ground in seconds, whereas planting and protecting a two-year-old bare-root sapling in grass can take five minutes or more—even in soil that's easy to dig.

The work may be slow, but it's also sociable. We usually work in teams of two or four—one or two people with shovels and the others with the plant bucket, plastic protectors, and coffee bags in a garden cart. We talk and work and work and talk: about our kids, the bald eagle passing overhead, the pileated woodpeckers yammering from the woods, the mouse tunnels in the grass.

And there is a lot to think about. Our boys and our friends in their twenties can think about the young forest their kids will run around in, the big trees they may see as old men and women. But if you're in your forties or fifties or older, these trees are not for you. Planting trees is a gentle way of acknowledging your mortality and celebrating it. The trees you are putting into the ground are a gift to the world. They are a thank-you present.

This is a good thing to think about at the end of the day. Gather your tools and your people together as the sun sets or the rain rolls in and look out at what you've done. There are little green sprigs scattered through the field. They will be a forest one day.

Blood, Sweat, Tears

When Susan's father was planting trees with his family, back in the late 1930s and early '40s, his mother had her very own planting system. As she finished putting a sapling in the ground, she would bend over it, wag a finger, and say, "Now, grow!"

And so they did. Forty-five years later, Carl built a house from some of those trees, cut during a thinning operation. More recently the Leopold family has donated posts, beams, and paneling to educational organizations in Wisconsin and beyond.

In the Tarboo Valley, we've looked at rings on recently harvested Douglas-firs and counted seventy on stumps that are 4 feet across. If our boys take care of themselves and are lucky, they may live to see trees that they planted get to that size. On the Olympic Peninsula, there are places where western redcedar trees are a thousand years old and where firs are more than 250 feet tall. At least a few of the little trees that we're planting may get that old and that large. But between planting two-year-old saplings and walking or harvesting a mature forest, there is work to be done.

~

One aspect of managing a newly reforested area is common to every restoration site, whether it's a prairie or a rainforest or a grove of mangroves: controlling invasives. Invasives are plants that are exotic—meaning not native to the site—and aggressive enough to outcompete the native vegetation. Only a small fraction of exotic plants—something like 10 to 15 percent—become invasive; why some do and some don't is a question biologists have been wrestling with for 150 years.

Initially, it seems paradoxical to observe exotic species outcompeting natives. After all, the natives have been living at the location for millennia and should be well adapted to local conditions. How do invasives outcompete them and take over? Sometimes the exotics use forms of chemical warfare that are new to the natives. Spotted knapweed, for example, emits an herbicide that native plants in North America have never been exposed to; the chemical it "sprays" may help it subdue susceptible competitors and take over sites. And unlike most plants, garlic mustard does not depend on fungi in the soil to supply its nitrogen and phosphorus in exchange for sugars. Instead, it secretes a fungicide that kills native soil fungi, causing their associated plants to weaken from malnutrition.

In many or most cases, though, the signature attribute of invasives is that they lack the herbivores and the pathogens—fungi, viruses, bacteria, and other disease-causing organisms—that native plants have to deal with. Think of the number of infectious agents and parasites you work to avoid, from colds and flus to mumps and food poisoning. The total is in the hundreds, and every organism on the planet has a similar array of viruses, bacteria, and other parasites that can infect it. To pathogens and parasites, a person or sword fern or cedar tree is a habitat: a paradise brimming with resources. Plants have insect enemies to worry about as well—some of which specialize in a single species, others that can eat a variety.

Trees and shrubs are constantly monitoring their own bodies and their surroundings for signs of danger. They can tell when their roots or leaves are being chewed, or when viruses have entered a wound and

are infecting their cells. They can even tell when nearby plants are being chewed by caterpillars or other herbivores; in response they manufacture proteins that clog the guts of herbivorous insects, and they sequester these poisons in their leaves, ready for the attack to begin. If they detect a viral infection, they wall off the site and kill the remaining cells within, wiping out the viruses in the process. And if they are attacked by caterpillars themselves, they may emit molecules that vaporize and are sensed by species of wasps called parasitoids. The wasps follow the scent trail to the plant and lay eggs inside the bodies of the caterpillars. The wasp larvae then eat the caterpillars from the inside out, killing them. In effect, the plants have called in close air support from waspy helicopters.

Imagine the competitive advantage for invasives, then, which don't have to expend all of this time and energy defending themselves. When researchers study plants that have become invasive by comparing their growth in a native habitat with their progress in a region where they've been introduced, the data usually support the predicted outcome: the invasive populations have fewer pests. They commit fewer resources to defense and more to getting big and gobbling up space. This is not ecological fair play. And there are consequences. A recent survey concluded that more than 20 percent of the endangered plant species in Canada are victimized by competition from invasives.

Introduced predators can be just as devastating, for the natives have no co-evolved defenses. Small, flightless birds called rails were no match for the rats introduced by humans to the Pacific Islands; David Steadman, curator of ornithology at the Florida Museum of Natural History, estimates that more than a thousand species of rails—essentially, at least one per island—were wiped out by rats who stowed away with the humans who colonized the Pacific.

Exotic diseases may be the most dangerous threat of all, though. Introduced strains of avian malaria wiped out dozens of native bird species in Hawaii; many of the survivors are reduced to relict populations at high elevation. Malaria is transmitted by mosquitoes, and the mountaintop habitats are too cold for mosquitoes to thrive. Similarly, the spread of West Nile virus is reducing bird populations in eastern North America; an introduced fungal pathogen is wiping out hundreds of species of frogs and salamanders all over the world; and white-nose syndrome is devastating North American bats.

Nonnative diseases are an issue for plants as well. Plant cells recognize pathogenic viruses and bacteria in much the same way as your immune system cells do. The process starts when proteins manufactured by our cells stick to specific parts of proteins from an invader. This binding event sets off alarm bells that lead to an effective defense response. But the proteins found on certain exotic fungi were never recognized by the sentinel proteins in American elms and American chestnuts because they were different from the proteins found on the native fungi. No binding occurred, no alarm bells rang, and two of the most important trees native to eastern North America were virtually wiped out. They never knew what hit them.

People have littered exotics all over the world. In Japan I've seen goldenrods from North America growing as invasives—each plant more than 6 feet high and featuring tough, woody stems. Hordes of them have taken over roadsides, forming dense monocultures—a far different situation from the scattered individuals found among dozens of other species in their native prairies. In Costa Rica, at the site where Carl led a successful effort to reforest abandoned pastures with native rainforest trees, the restoration crew had to battle grasses imported from Africa.

At Tarboo Creek the bad actors are holly trees, reed canary grass, and Himalayan and Eurasian blackberry bushes. Holly grows in the woods, reed canary flourishes on wet soils, and the blackberries are everywhere. Each of us has it in for one of them. Our older son cuts holly on sight; Susan can't pass a clump of reed canary without stomping on it or trying to yank it out; I can machete or dig out blackberry for hours. Our younger son is a gentle soul. He lives and lets live—but only because his least-favorite invasive, Scotch broom, is rare at our place.

The reed canary and blackberry should eventually disappear as the trees grow and shade them out. But we will never be rid of the holly at Tarboo because it's shade tolerant—it can grow under a forest canopy. We also have neighbors who own a copse of mature holly trees, each of which bears bushels of bright red fruit in winter. The berries make wonderful holiday decorations and are a favorite with the robins. The birds eat the berries and scatter them around our woods along with little packets of fertilizer—sowing the seeds of destruction.

Invasives are a bad business. Cheatgrass has forever altered the short-grass prairies and sagelands of the Rockies and Intermountain West; the economic losses to the ranching industry are incalculable. Kudzu has formed impenetrable monocultures in abandoned farmlands of the American South that otherwise would have succeeded to productive forest.

~

When biologists catalog threatened species and analyze the causes of decline, they find a strong historical shift: until about 1950, most extinctions and other declines were caused by overhunting. Overfishing is still a concern for many marine species, and overhunting is causing problems for certain terrestrial mammals, like elephants, and birds in selected areas. But the general situation has changed in a fundamental way: the overwhelming majority of current and projected problems are due to habitat loss and invasive species.

Although invasives can be deadly, ecologists and restorationists are beginning to realize that they are also a fact of life. For better or for worse, humans have changed the geographic ranges of more species faster and more broadly than at any previous time since life began.

We can't go back and undo the changes or the damage. Even if we had the time and the money, invasives are too well entrenched to be eliminated entirely. The only real solution is for evolution by natural selection to run its course. If a lucky mutation in an American elm allows it to recognize and respond to its introduced fungus, or if individuals from native fungal species experience chance mutations that make them immune to garlic mustard's poison, they will be fruitful and multiply. At some point, a virus native to North America will have a mutation that allows it to infect cells in the stems of reed canary grass, smiting it. When some beetle sustains a mutation that allows it to digest Himalayan blackberry sap, that fortunate insect will inherit the earth, or at least a small part of it. I hope I live to see the day.

In the meantime, restorationists cope. Occasionally I work through a blackberry patch with nippers in one hand to cut the stems and an herbicide-laden brush in the other to wipe the stumps. But more frequently we cut and recut. We are trying to keep the invasives down to a dull roar, buying time and space for the natives that we are introducing to get their feet on the ground and overgrow the exotics. We play the role of umpire, trying to level the ecological playing field.

Planting native species is a way to practice poetry, but controlling invasives is all prose. The work makes you sweat; the thorns and prickles draw blood; their persistence brings you to tears.

~

The first restoration project Susan and I worked on together was near the Leopold Shack—the abandoned farm in central Wisconsin that Carl and his parents and siblings had transformed a generation before. On our walks around the place we found a little postage stamp of native prairie: a 4-acre site where a few native grasses and wildflowers were struggling to hold on, and where some swamp white oak trees were growing. It was a relict savanna—a plant community made up of a prairie understory with scattered trees.

The neighborhood where I grew up in southern Wisconsin and the campus where I was a college student in southern Minnesota had been covered with burr oak savanna when Europeans arrived. Burr oaks are stout, slow growing, thick barked, and deep rooted. All of these characteristics helped the trees withstand intermittent droughts and the fires set by native people from the Sioux, Fox, and Winnebago nations to kill invading trees and encourage grass for deer, bison, and elk. Because they grew in the open, the fire-resistant burr oaks developed wide, craggy, umbrella-like crowns. The combination of burr oaks and prairie grasses created parklands in midwestern North America akin to the Maasai Mara National Reserve in east Africa.

People respond to these environments instinctively. As our eyes sweep across a landscape dominated by savanna, it's as if our subconscious recites, "Here is pasture for my cattle; wood for my hearth and home. The terrain is easy to travel and teeming with game. There is nowhere for my enemies to hide. The ponds and rivers have fresh, clear water, teeming with fish. Here I will live."

We evolved in savannas and are hardwired to love them. That's why landscape architects are always trying to reproduce them, using exotic trees and mowed grasses in corporate and college campuses. I even wonder sometimes about the savanna-like appearance of strip developments in suburban settings. Instead of grassland, there are expanses of concrete. Instead of acacia trees, there are scattered, spindly signs advertising gas stations or franchise restaurants.

When Susan and I found this bit of ragtag but actual savanna near the Leopold Shack, it was being invaded by a shrub called prickly ash. Although prickly ash is native to North America and isn't considered an invasive, it started to take over vast swaths of the Wisconsin River floodplain after white settlers pushed the native people out. The whites didn't follow the traditional practice of setting fires to clear out woody shrubs and young trees and encourage the prairie flora, and lack of fire led to a dramatic increase in prickly ash. Its takeover of the savanna wasn't quite complete, however. Under and between the dense thickets of prickly ash stems, we found wisps of cordgrass, bluejoint grass, and other native species. But what made the place special was the swamp white oak trees.

The burr oak savanna that once grew in my boyhood neighborhood is rare now—vast areas have been converted to farms and suburbs, while other sites succeeded to continuous forest after Europeans arrived and controlled wildfires—but swamp white oak savanna is almost nonexistent. Swamp white oaks have such a limited geographic range that the habitat type was never common, and today this version of savanna is essentially extinct. We thought of Susan's grandfather, who justified his restoration efforts in the 1930s by observing, "The first step in intelligent tinkering is to save all the parts." Swamp white oak savanna was a part that had almost been lost, so we wanted to save the little patch near the Leopold Shack.

In this case, the restoration plan was straightforward. The first step was to cut the prickly ash out; the second was to start a regular program of prescribed burns. The fires would keep the prickly ash down and out and encourage natives to fill the vacated space. Once the burning regimen was under way, we could seed in new prairie species to increase diversity.

Step one took a day—a long day. We recruited a gang of friends and relatives for a work party one hot summer Saturday. A cousin of Susan's and I each made a vanguard, running handheld brush cutters—long bars with a two-cycle engine on one end cabled to a circular saw blade on the other end—into the tangles of prickly ash. The rest of the crew worked behind us. As we scythed back and forth with the saws, the crew bucket-brigaded the cut stems out and threw them into great stacks—thorny versions of Monet's haymows.

It was brutal work. Prickly ash is weaponized with thorns, and even with long sleeves and leather gloves, Susan's cousin and I were scratched and clawed and bleeding. At lunch, when we'd completed about half of the cutting, he collapsed against a grassy bank. His shirt was streaked with mud and soaked with sweat; his horn-rimmed glasses had been knocked askew; his prickly ash gashes were glowing red. He looked at me and pleaded, "Now, tell me again: *Why* are we doing this?"

I didn't have much of a response then, but a year later, we did. Friends helped us burn the patch early in the autumn after the work party, and the following summer, scattered among the charred and rotting prickly ash stumps, Susan and I found a spectacular show of bottle gentian—an uncommon forb with a flower so purple it should be worn only by royalty. The gentians hadn't been able to bloom there for decades. Susan's Savanna is now one of the pearls of the Leopold Memorial Reserve.

⁓

In the Tarboo Valley, the prickly ash equivalents are the exotic shrubs called Himalayan and Eurasian blackberry. Peter Bahls has a name for blackberry removal: "The worst." He hires crews of twenty-somethings to cut it and then dig it out in areas where plant-a-thons are scheduled for the following winter. The NWI crew members end up as sweaty, scratched, haggard, and muddy as Susan's cousin and I were. Indeed, a family friend who did blackberry removal in the Seattle parks system told a story that captured the toll blackberry can take. She was in Seattle's University District after work one evening, waiting for a bus home, and got panhandled from behind by a street kid. When she turned to face him, he backed away, horrified. He took in her blackberry scratches, dirty face, and disheveled clothing and said, "Oh, I'm so sorry, man. Hey, good luck."

When I spend a weekend cutting blackberry at our place, it's often in thickets where the blackberry canes are growing over the tops of the other shrubs, 8 feet in the air. I show up at work on Monday morning with scratches on my forehead and scalp. I get a few "What happened to you?" queries, but there isn't much to say. Invasive control is down and dirty.

~

On every trip to Tarboo Creek, our older son reminds us to do something constructive and something destructive. If we rip into a blackberry patch in the morning, we plant trees or put up a wood duck box or take a long walk to check on saplings in the afternoon. The walks through the restoration area are especially rewarding. You are visiting loved ones with your loved ones—checking on their health and well-being.

Ten-plus growing seasons after the initial plant-a-thon at our place, many of the trees in the grassy old pasture have just begun to grow in earnest. Their root systems have finally gotten big enough to compete effectively with the overlying grass net. For years their trunks and branches had a cramped and stunted look, as if they were hunched over in pain. Some were barely 2 feet tall five years after planting, not even double the size they were on planting day. But closer to the creek, where the grass was partially shaded by the alders that remained near the stream channel, and where the low-lying soils don't dry as thoroughly in late summer, the trees began to flourish almost immediately. Within a few years their leaders were casting upward 3 feet a year; within the decade the tree canopy had closed and the creek was shaded and cooled.

Years later, virtually all of the trees that survived their first few years in the pasture finally began to get broad enough to start shading out the grass below. It will still be many years before the canopy closes in the pasture and the grasses begin to be replaced with shade-loving forest herbs, shrubs, and ferns. But now, the seedlings' roots are thick and broad enough to provide a strong foundation and win the competition for soil nutrients and water. We no longer have to maintain the restoration; it maintains itself. At the tipping point when the saplings begin to win and the grasses and invasives begin to lose, restorationists say, "The trees are free to grow."

I like that phrase. It's a goal in land stewardship that's parallel to the goal Susan and I had as parents. We wanted to raise our boys, like these trees, so they got to that tipping point and no longer needed our care. Thereafter, they were strong and independent. They were free to grow.

～

As we walk through the restoration in the evening, checking on the trees, Susan has the habit of touching each one as we go by. She isn't even aware; she just touches them. And we comment as we pass: "Wow, this Doug-fir is taking off. . . . Too bad—this cedar just died. . . . We should add a couple hemlocks here. . . . I wonder if grand firs would work in this area. . . . Look where a black-tailed buck has been jousting with this willow, rubbing velvet off. . . . Weevils are killing the tips of this spruce."

We were doing this once when Susan's father was visiting. Carl was eighty-four at the time, and as we weaved around the saplings, talking and commenting in the sunset light, he exclaimed to no one in particular: "This is just what we did sixty years ago. It's *just* what we did."

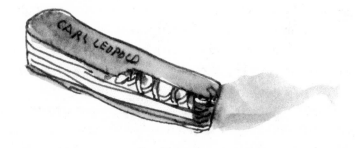

A Working Forest

~~~~~~~~~~~~~~~~~~~~~~~~~~~~~~~~~~~~

Carl Leopold died on November 18, 2009. He'd spent more than sixty years working as a research scientist—a plant physiologist who studied how plant hormones function and how seeds survive extreme drying and long periods of dormancy. He planted trees wherever he lived and was a founding director of the Tropical Forestry Initiative, the Finger Lakes Land Trust, and the Aldo Leopold Foundation. At age eighty-nine he was still riding a moped to the office; he worked the day he died.

As a memorial to him we bought a tract of forest near our place at Tarboo Creek. Most of the woods had been clear-cut in 1965 but the corporation that owned it had left large numbers of cedar saplings standing—probably with the idea of selling a "green" clear-cut for development. The development never materialized, so third-growth timber grew. The company pulled out of the Pacific Northwest in the 1970s and sold the parcel to a Danish investor who had been buying up timberland in Oregon and Washington for decades. At the time we bought the land, he was planning to clear-cut the forest, subdivide, and sell lots for development.

Second home and retirement home development has emerged as one of the great threats to open space in the United States. Puget Sound is ringed with view homes, and throughout the United States suburban sprawl has converted millions of acres of working timberland and farms, burned up trillions of dollars in gas, and wasted billions of hours in commuting time—when people could've been working or playing with their kids. We wanted the Carl Leopold Forest to offer an alternative.

Intense pressure on farms and forests is building all over the world. When Carl was a teenager, there were about 127 million people in the United States and just shy of 2.2 billion in the world. As I write this, there are almost 324 million people in the United States and more than 7.3 billion on the planet. In 2050, when today's teenagers will be in their fifties, researchers estimate that world population will be about 9.5 billion. Those projections assume that fertility rates—the average number of children a woman has during her lifetime—will continue to drop from the historical highs that occurred between 1965 and 1970. If fertility rates don't drop as dramatically as the optimistic projection, the best analyses suggest that world population will be closer to 11 billion by 2050. If this scenario plays out, there will be more than 16 billion people on the planet in the year 2100, when the children of today's teenagers will be elderly. And if the fertility rates observed today continue into the future—that is, if there's no further reduction in the global average number of children per woman—there will be more than 28 billion people on Earth in 2100.

We worry now about traffic, the price of food and gas, and whether our kids will be able to afford a house. It's hard to imagine what the world of 2050 will be like with an additional 2.2 billion or 3.7 billion people. Faced with projections like this, thinking globally can be overwhelming. But acting locally can still be inspiring.

〜

In the main part of Carl's forest, the little cedar whips that were spared in the 1965 cut are now 100 feet tall and 3 to 4 feet across where their flaring stems plunge into the ground. For reasons that are a complete mystery to us, there is also a 7-acre stand of big Douglas-fir trees— some of them more than a century old, 175 feet tall, and 4 feet in diameter at breast height. Neither we nor a forester friend can figure out how they avoided getting cut decades ago. The tract also hosts three tributaries to Tarboo Creek that flow for most or all of the year. Two have carved dramatic canyons into the hillside and the other splays into four smaller channels, like fingers from a palm. The ridges between these fingers are littered with huge rotting logs left over from the old-growth days.

Carl's forest is a memorial, but it is not a preserve. It's working land—a working forest. In addition to producing sustainable harvests of wood products, we hope to reproduce the four structural characteristics of an old-growth forest.

First, downed woody debris on the forest floor. In old-growth forests, up to 10 percent of the soil surface can be covered with downed wood. The largest constituents are tree trunks that are slowly rotting away. As termites, ants, fungi, and bacteria recycle nutrients and soften the wood, the logs gain the ability to soak up winter rain and store it, spongelike. The moisture helps to keep the roots of nearby living trees moist in late summer and supports the growth of tree seedlings down their length—the nurse-log phenomenon.

Second, standing dead snags. These are stately columns—the ruins of majestic old trees. Snags go through five stages of decay over a span of two hundred years, a little like Shakespeare's seven stages of life. At an advanced age they have sloughed their bark and branches and stand naked to the world, their tops broken off by windstorms. They are long dead but also full of life, for they're riddled with cavities that provide warm, dry, secure homes for flying squirrels, bats, owls, and woodpeckers; if they hollow out at the base they also offer denning sites for bears. Almost a hundred species of birds and mammals in

the Pacific Northwest use snags; more than half of those species can't live without them.

Third, some two-hundred-year-old trees. These are the giants we all associate with old growth—great beasts with limbless stems rising hundreds of feet above the forest floor. Their lower branches can be 8 inches or more in diameter and are invariably draped in air plants, or epiphytes. In the tropics, trees are festooned with orchids and bromeliads; in the Pacific Northwest, epiphytes range from pillow-like tufts of moss and sprays of licorice fern to rubbery, leaf-shaped lichens. Near the coast, the broadest and mossiest of those branches are home to one of the last North American bird nests to be discovered and described—that of a little seagoing bird called the marbled murrelet. The big trees are important to people, too. An acquaintance named Laura Solomon from the Lummi Nation near Bellingham, Washington, told me about a stand of old-growth Douglas-fir that she'd been visiting for years but that had been recently logged. "These are our churches," she said. "Cutting that forest was like destroying a cathedral."

Fourth, complex structure. By this, scientists mean that from the forest floor to the tips of the largest, canopy-forming trees, there are multiple layers of vegetation. Below the uppermost layer, or canopy, is a stratum consisting of large trees that are still growing toward the tallest "dominants." Below them are an array of small trees, then shrubs, and finally ferns and herbaceous plants on the forest floor. The layers make efficient use of the incoming sunlight, and their presence means that food and places to live are available at a wide array of heights above the ground.

This is the kind of forest that Lewis and Clark found and that Chief Seattle grew up in. It is also one of the most effective habitats in the world in terms of sequestering carbon.

I'll never forget the first time we took Carl Leopold to see old-growth trees. He looked up in awe at a monumental Sitka spruce and breathed, "Think of all the fixed CO_2." When that spruce took up carbon dioxide molecules from the air, fixed them into a sugar molecule during photosynthesis, and used that sugar to manufacture the cellulose and lignin that stiffened a new cell wall in its trunk, it added to what scientists call a carbon sink. Every one of the carbon atoms in that massive tree will stay in place for hundreds of years while it is living and some additional hundreds of years as the trunk and roots slowly decay. Finally, each carbon atom will be released into the atmosphere in the form of CO_2 as bacteria, fungi, and other decomposers exhale.

Recent analyses have debunked the long-standing view that old forests stop accumulating carbon. The conventional wisdom was that as growth slows in stands of old trees, just as much carbon dioxide is released to the atmosphere from decaying dead wood as enters living trees via photosynthesis—meaning that carbon use and carbon release are in a steady state. But it turns out that this view was based on the results of a single study published in 1969. Research over the past forty years in hundreds of sites in temperate and northern latitudes has shown that older forests continue a net accumulation of carbon into old age—up to eight hundred years—and that old growth in the Pacific Northwest contains the most wood, and thus the most carbon, of any forest type measured to date: up to 1,600 cubic meters per hectare.

Ironically, young forests can act as net sources of carbon dioxide, even though the trees may be growing much more rapidly than old-growth trees. A large amount of the carbon in mature forests is stored in the soil as humus or in tree roots. If the mature forest is wiped out by fire or logging and roots begin to decay, the amount of carbon released from the underground reservoir can exceed the

amount taken up by the young trees. When this occurs, converting older forests to younger forests produces a net increase in atmospheric carbon dioxide and thus heat-trapping gases. It has a positive feedback effect on global warming.

We've already started to let windfall trees in Carl's forest stay down and insect-killed trees stay up. Friends from Europe are perplexed at this—their forests are clean. Compared to woods in the Old World or an industrial tree farm in the Pacific Northwest, Carl's forest looks a mess.

Other habitats can be effective carbon sinks as well, but most of them are unpleasant places to spend time unless you're an anaerobic bacterium. In most areas that function as carbon sinks, organic matter builds up and sequesters carbon because decay is slow—the carbon present in cellulose, lignin, sugars, and other organic compounds isn't being released back into the atmosphere as CO_2 very quickly. Decay is slow in habitats that are cold, and extremely slow when oxygen is scarce or absent. Today, the major carbon sinks include oxygen-free waters in the bottom of the ocean, peats that form in the Arctic or in bogs, and anaerobic mucks and muds that reek of hydrogen sulfide. These aren't places you want to go on a first date. A wildflower-strewn prairie—where carbon is being added to the soil by roots growing 10 or even 20 feet underground—or an old-growth forest is an attractive alternative if you want to store your carbon and enjoy it, too.

～

When Susan and I moved to the Northwest from Wisconsin in 1985, we arrived to find a war in the woods. Old-growth forests in northern California, Oregon, Washington, and British Columbia had been cut since the mid-to-late 1800s, starting at sea level and working up the slopes of the Cascade and Olympic mountains. By the 1980s, only about 10 percent of the original stands remained in Washington and Oregon, and almost all of the remaining old growth was in national forests. Large-scale replanting of trees had started only in the 1940s and '50s; the industry's plan was to cut the remaining old growth in the 1980s and '90s and buy time for the plantation trees to mature.

This plan ran into a buzz saw of lawsuits filed by environmental organizations. The suits claimed that applications to cut the remaining old growth violated the U.S. Endangered Species Act because they put the northern spotted owl—a bird that relies on old-growth habitat—at risk of further decline. The legal action was successful; it virtually halted the cutting of old growth on public land.

The economic impact was tremendous—at the time, forestry was the Northwest's largest industry. Job losses were severe because harvesting old growth is labor intensive. The trees have to be cut by a feller with a chainsaw, then limbed and sectioned by hand before they can be loaded onto trucks and transported to the mill. Those jobs disappeared. The old mills, which were set up to handle the gigantic old-growth logs, shut down—sooner than the industry had planned.

When the war was over, the industry had changed dramatically. The inevitable transition to processing second- and third-growth timber happened sooner rather than later. Mills retooled to handle small logs from forty-to-sixty-year-old trees grown on plantations and were reengineered to be almost completely automated. Cutting small second-growth trees also required much less handwork than cutting old growth. In the Tarboo Creek watershed, we've seen 60-acre clear-cuts run by two people: one running a feller-buncher —a cutting-and-limbing machine that lumbers along on tractor treads—and a second running a loader with a gigantic claw for picking up trunks and piling them into stacks that can be loaded onto waiting trucks. The pattern of land ownership changed, too. Many of the big private firms turned at least part of their attention to real estate development, selling land with access to power lines for second homes, retirement homes, and hobby farms, or even entire planned exurban communities. And recently, more and more timberland has been bought up by real estate investment trusts, or REITs—groups that are betting on a growing human population to furnish long-term demand for forest products. They hold wood-growing land as an asset.

In a sense, the forest products industry in the Northwest has followed a trajectory similar to agriculture in other regions of the United States. Large forest-product companies or REITs manage their land to maximize income and productivity—with clear-cuts scheduled on forty-to-fifty-year or even shorter rotations, then replanted to fast-growing cultivars of Douglas-fir. These outfits grow small trees and sell 2-by-6 studs, plywood, chips used in papermaking, and other large-volume, low-priced commodities. It's a cropping system. Because they own and process their own monocultures, their business model echoes the elements of big agriculture, where specialist corn and soybean growers sell to a few commodity-processing and marketing behemoths, which market to industrial-scale meat producers or fast-food companies buying in enormous volume.

But small landowners, who grow bigger trees and may harvest by selective thinning rather than clear-cutting, also have a role to play. These are family-owned businesses that grow six or more different species and produce bigger logs used in more specialized products like posts, beams, cedar utility poles, flooring, paneling, or knot-free, furniture-quality wood. These small, family-run operations are akin to the new farms that are cropping up near our place in Washington's Jefferson County. There, a cadre of young and well-educated farmers, orchardists, and food processors are starting successful businesses. They practice a less capital- and more labor-intensive style of farming and produce high-value specialty products: fresh vegetables and organic grains, cheese, free-range chicken and eggs, restaurant-quality beef and pork, hard ciders and fruit brandies, and

organic fruits and berries. Most are vertically integrated businesses: they do their own marketing and sell directly to people who want to eat locally produced, high-quality food.

In essence, the ownership of farmland and forests in the United States has followed the old Hamilton-Jefferson split: a numerically dominant, mercantilist economy is driven by financiers and large corporations, and a relatively tiny agrarian society is run by small family farmers. For more than two hundred years, the Hamiltonians have won almost all of the political and economic battles; government agencies and tax policies are organized around their needs. But the Jeffersonians—the young, innovative farmers and foresters—are still here working the land. And like most dichotomies, the big ag and small ag divide isn't exclusive. Both have a place in a healthy rural economy and landscape.

~

Our goal is to manage Carl's forest for the long term and produce a steady stream of high-quality wood products. One of our first management efforts toward this objective focused on a 40-acre tract adjacent to the north edge of the older forest. This is a stand on an east-facing slope, bisected by a tributary to Tarboo Creek, that was clear-cut in 1997 and then put up for sale.

At fifteen to twenty years of age, regrowing trees start to form a closed canopy and can benefit enormously from what foresters call a precommercial thin. The adjective means that the thinned trees aren't sold but simply dropped and left to rot and add organic material to the soil. Our precommercial thin had two goals: giving the most vigorous trees room to grow and maintaining or increasing species diversity. As we moved through this young stand, clad in helmets, Kevlar chaps to prevent the signature chainsaw wound (a deep gash across the thigh), gloves, and ear and eye protection, we were trying to achieve an overall, average spacing between trees based on research regarding how stands respond to different severities of thinning. But after that, we made decisions on the fly. Which stems would live and which would die? Red alder matures much earlier than Douglas-fir, which matures much earlier than western redcedar. In some areas we cut to clump individuals from the same species so they could be harvested together; in others, we actively favored the rarer species to increase overall diversity. We dropped trees that had lost their leaders and were growing twinned stems, and we eliminated trunks that were misshapen—unless the art in their knobby projections and sweeping curves spoke to us.

The work is gritty and exhausting but exhilarating: you leave behind a forest with light that reaches the soil surface, ready to awaken a dormant understory layer of voluptuous ferns and berry-bearing shrubs, and gaps in the canopy layer, soon to be filled by grateful trees growing in ferocious profusion.

Once the initial thinning is done, we touch the forest again, with handsaws. This is a pruning pass: cutting limbs flush to the trunk, initially up as high as a person can reach. Pruning transforms a young stand because you can suddenly see through it as tens of thousands of shade-killed limbs drop to the ground. Within a few years of pruning, the trunks will grow over each wound, creating a smooth outer surface. The wood that grows thereafter will be clear—meaning free of knots. Logs with branches, in contrast, produce one of two types of knotty wood. Lumber with the flaws known as knotholes is sawed from trunks that have grown over dead limbs. The wood from the dead branches tends to shrink and drop out of the board, leaving a hole. Lumber with tight knots is sawed from trunks that have grown over live limbs, trapping the tissue growing out from the main stem. But clear, knotless lumber is sawed from limbless trunks. It commands a premium price due to its beauty and strength.

As the years progress, we'll enter the forest again with pole saws to do lifts—removing additional whorls of branches as the trees get taller and taller. With our tools and our sweat, we simply accelerate the process of creating tall, straight, limbless boles that would otherwise happen naturally.

～

Thinning and pruning have spiritual benefits too, for there is no better way to learn a patch of land than to work on it. We've found warrens dug by mountain beavers—pug-size offshoots from the base of the rodent evolutionary tree that live only in the Cascade and Olympic mountains. I've watched an Anna's hummingbird follow Susan when she's wearing her favorite red sweater and virtually land on her shoulder, mistaking her for a blossom. On occasion we see black bear sign: a pile of scat, old rotting logs and stumps that have been torn apart in search of grubs and ants, and Douglas-fir trees with bark ripped off. Bears will bite the bark of young trees in late spring, when the sap is running at its peak, and peel it up the trunk. The effort exposes the sugary cambium layer, which they eat. Bears will even return in subsequent springs to harvest the tissue next to the wound, now swollen with sugars the tree has sent to speed the repair. Although most trees recover from the assaults, some are girdled and die, forming the forest's first standing dead snags.

Working in the woods has another benefit as well. In Carl's forest, the work of thinning and pruning is a family affair. We toil together. The discussions and decisions, the scratches and bleeding, tie us to the land and to each other.

Carl's forest should become more and more valuable over time, producing a sustainable harvest of high-quality logs for market and taking on the four characteristics of an old-growth stand. I can't think of a better way to honor a man who endured the Great Depression, fought a war, raised a beautiful family, loved music, contributed to science, saved land, and befriended people from all walks of life.

Damnation

~~~~~~~~~~

When the trees we'd planted along the remeandered sections of creek were just a few years old, I noticed one or two disappearing at a time—usually in the fall. They were western redcedars in every case, and they were cut just above the plastic plant protector tube. I was mystified—the cut point was too smooth to be deer work. Deer have teeth at the front of their top jaw but the matching surface of their bottom jaw is toothless—it's just a bony ridge. Instead of cutting, they have to tear stems and twigs as they browse, leaving a ragged edge.

I wondered for a time about mountain beavers—the overgrown hamsters we find in the younger stands in Carl's forest. Their common name is a misnomer, as they aren't particularly closely related to regular North American beavers. Mountain beavers live in uplands, make mazes of underground burrows, and eat just about any type of plant; they're particularly fond of ferns. But the little cedar trees had gone missing in the floodplain, next to the creek channel, and there was no sign of soil disturbance save for an occasional molehill.

The answer came to me at last in the form of a stick. An alder branch with the bark chewed off floated on the creek toward me. I fished it out of the stream and looked at it. There were neat, paired grooves circling the stem, making a series of tiny ridges—a miniature version of a Latino band's percussion instrument, a guiro—and the ends were chewed into triangular points. This was incontrovertible evidence of *Castor canadensis*. We had a North American beaver on our hands. Up until then, the only beaver activity I'd been aware of was well over a mile away from our remeanders.

At first we didn't make too much of this. According to our thinking at the time, a beaver might take some willow and alder stems, but plenty were available. During the remeandering operation we'd secured much of the coir matting along the bank with willow stakes, and many of them had sprouted. Alder was seeding in on its own, and the alder pull-ups we'd planted were taking off—growing like weeds. So we had lots of beaver food at the ready. Noticing some beaver activity was like having someone who plays loud music on Saturday nights move in a couple of doors down. You remark on it, but it's nothing to get that concerned about.

Then one autumn afternoon we found a much bigger cedar sapling cut. Of the thousands of trees we'd planted, this one had grown the most—it was in the pride and joy category. But its 3-inch-wide trunk was cut at the telltale location, just above the plant protector, and the 8-foot-tall stem lay on the ground, stripped of its leaves and branches. With that, the truce was shattered. The beavers were ignoring the willows and alders. They preferred our cedars, and the deep shade and long lifespan of western redcedars were key to the stream's future. We had to do something.

～

Friends and neighbors had advice: get rid of them. Almost universally we heard, "You don't want beavers. They're so destructive."

This advice didn't sit well. First, there was something ironic about the emotional responses. Humans have dammed the Columbia and its tributaries more than sixty times, along with the Nile, the Yangtze, and the Colorado, to name a few. We have channeled the Mississippi and the Thames and destroyed most of the world's prairies and old-growth forests. And we call beavers destructive?

Second, neither of the removal options made sense. Urban friends advocated livetrapping the beavers and moving them to a new location. This is a popular thing to do with nuisance squirrels and raccoons in cities; it makes people feel humane. The problem is that dropping animals off in a new location is like knocking on the door of someone you've never met, introducing some strangers that you don't want living in your own neighborhood, and telling the homeowner that the strangers are going to move in. The newcomers are not likely to be welcomed. The transplanters may feel humane and congratulate themselves, but the transplantees are likely to die slowly and painfully, of starvation or wounds from confrontations with the local residents.

The other option, killing the beavers outright, is neither amoral nor unethical. We kill things all the time, from swatting mosquitoes to harvesting the plants and animals we eat every day. There are, however, two strong counterarguments to killing beavers. The first is a practical issue of time and energy. Trapping beavers puts you on a slippery slope, because the individuals you remove will be replaced by other beavers—usually sooner rather than later. Unless we intended to remove beavers from the entire watershed, we'd be committing ourselves to a lifetime of feuding. Second, and much more important, beavers can be critical to the success of a salmon stream restoration. They are what biologists call ecosystem engineers. The problem was that we needed to direct their energy—away from eating trees that also were critical to the success of the salmon stream restoration.

～

If your goal is to live with the land instead of just on it, you have to accept the organisms that live there more or less on their own terms. In our neighborhood, this might mean not digging a trout pond when river otters are common, or not mowing an acre of lawn and then complaining about molehills in it. It might even mean not moving to a rural area at all—but just visiting.

We'd faced this issue before in another context. Along our section of creek, Bob Harrison used the excavator to plant what we call eagle poles—large logs that stick up a couple of stories aboveground. At the NWI land just downstream from us, there are so many of these artificial snags that Peter Bahls calls it Woodhenge. Both restoration sites have large logs scattered on the ground as well, artfully arranged by Bob's excavator. The idea here was to kick-start a restoration with two of the four characteristics of an old-growth forest. Two centuries before big trees will have grown and decades before a complex forest structure can develop, you can have standing dead snags and downed wood simply by adding them. So installing snags and scattering logs on the ground were on the task list for Bob and the Hitachi.

The first fall that salmon ran through the new meanders at our place, though, we found skeletons and shreds of flesh at the feet of some of the poles. Eagles were using the poles as hunting perches: monitoring the creek, stooping on fish, and returning to eat their kills. This drove Bob to distraction. "We're doing all this work to make the stream perfect for salmon, then we plant these damn poles to make it all nice for the eagles to eat the fish!" The only response was to nod. This is what life is like when you're part of an ecosystem.

~

Beavers build dams to create safe zones. On land, they are almost defenseless against attacks by coyotes, cougars, wolves, or bears. But in the water, they're safe. They are strong swimmers and have fortresslike lodges they can retreat to. If a dam is large enough to pond water into a small lake, beavers will build a lodge in the middle

so it's surrounded by a moat. If the site is far enough north for the water to freeze over in winter, they will also invest in a food cache—a mound of freshly cut trunks and branches piled on the lake bottom—in the vicinity of the lodge. When the surface freezes inches or even feet thick, they can leave their lodge whenever they want and dive down to retrieve a snack. But along smaller streams they build a bank lodge—digging out an entrance tunnel and an exit tunnel so they can't be trapped inside. The tunnels connect with a snug chamber where the family can rest in peace. Building a dam across a small stream and ponding the water behind it can make it possible to have an underwater, ultra-secure entrance and exit for the bank lodge.

Although beavers drop trees to provide material for making dams, their primary goal in logging is to eat. Their cheeks look puffy because they are distended by massive jaw muscles; their paired front incisors grow continuously and are sharp and hard enough to take down maples and other hardwoods. To cut a tree, they move around the trunk making a V-shaped cut, similar to what a woodsman would do with an ax. When beavers cleared out a grove of vine maples at our place, they left a group of thirty-two gnawed-off stems clumped in the space of a few square yards, each sharpened to a point like a punji stick. Once a stem is down, they buck it up the way a human logger would—in this case, cutting it into segments that can be dragged to the water and eaten in safety.

In summer, beavers eat the leaves of the deciduous trees and shrubs they harvest; all year round they eat conifer needles. The staple of the beaver diet, though, is cambium and vascular tissues. These cell layers, found in the outermost part of a trunk or branch, consist of three major components: stem cells that divide to form new wood, the structures that form a tree's circulatory system, and bark. As cells in the stem-cell layer divide, the daughter cells move to the inside or outside, steadily increasing the girth of trunks and roots. Some of these cells differentiate to form the vascular system tissues called xylem and phloem—groups of specialized cells that transport water and sap, respectively. Together, the cambium and vascular tissues form a thin, living layer of cells just outside the stem's woody core and just inside the outer bark.

The wood and bark on either side of the cambium layer consist of dead cells dominated by cellulose and lignin—two molecules that mammals can't digest. But the cambium and vascular tissues are moist and nutrient rich. They are also sweetened by sugars—molecules that are manufactured by photosynthesis and then transported back and forth between mature or growing leaves and storage areas in the roots. You've tasted the sugars in phloem sap if you've ever eaten maple syrup; around the world, almost all aboriginal people will list cambium and vascular tissue as traditional food sources.

The stuff is so pitchy, though, that it's almost always eaten as a last resort—a starvation food.

You can tell where beavers have been feeding by spotting branches and logs that are bright from a recent barking: sticks that have had their cambium chewed off lie scattered about like bleached bones. But beavers leave other sign as well. You can look for scent mounds along lake or stream banks—*No trespassing* signs for other beavers to note—or tracks in mud at the water's edge. Along both streams and ponds, you'll find skid roads where beavers have hauled themselves up and out of the water, humped across land to a logging site, and then dragged tree sections back. In marshes or other shallow wetlands, it's common to find networks of deep ditches, dug to provide a swimmable and safe road network. And you should listen, too, for the slap of a tail hitting the water surface: a warning signal if you happen to surprise a family that's out and active. It's extremely rare to find beaver scat, but it's not hard to recognize if you're lucky enough to run across some: it looks like sawdust.

～

In 2006, a group of paleontologists working in China reported that they'd found the almost-half-meter-long skeleton of a furred animal with a beaverlike tail and evidence of webbed hind feet, in rocks that were laid down 165 million years ago. But neither the skeleton nor the dentition were rodentlike, and its teeth suggested that it was mostly carnivorous, like a river otter. The best evidence available suggests that its close relatives belong to an extinct order of early mammals called the docodontans. The tail, streamlined body, and webbed feet are examples of convergent evolution—adaptations that have arisen in groups as diverse as today's platypus, otters, and beavers. Platypus lay eggs and are in the mammal lineage called monotremes; otters are in the group called Carnivora and are especially closely related to weasels; beavers are rodents, so have mice and rats as close cousins.

The discovery of this Jurassic animal popped a hole in the idea that mammals were exclusively small bodied and were bit players ecologically until the demise of the dinosaurs. It also suggests that the beaver way of life—having a body specialized for swimming and for eating resources available in streams and wetlands—has something going for it. Aquatic mammals have a long and storied history.

Certainly, beavers have been living in Tarboo Creek for as long as the stream has existed. Even after a century and a half of intensive disturbance, there are still flattened, peaty sites in the valley descended from long-filled beaver ponds. Directly north of Tarboo, the Chimacum Creek watershed was a continuous series of beaver dams and ponds when whites arrived; just to the north and east, settlers called another watershed Beaver Valley.

Beavers frequently dam the small streams that most salmon prefer for nesting. But any and every salmon biologist will also tell you that beavers are among the best things that can happen to those small streams. Instead of having their passage blocked, the fish almost always find a way through, up and over, or around beaver dams. We've watched several different tactics. The first is the most direct. If the barrier is leaky enough, the fish will simply wriggle through.

The second may be the most dramatic. In a classic beaver dam, sticks on the downstream side form a wall that slopes at 30 to 45 degrees, strengthening the structure. Coho salmon are accomplished jumpers and will throw themselves on this sloping face and thrash, throwing their bodies up until they flop over the top. Or if the stream is at flood stage, small breaks may open in the top of the dam, forming channels that cascade down and that fish swim up.

Third and last is an outmaneuvering approach. If the dam is too high or too steep for the up-and-over strategy to work, the fish will try to outflank it. In low-gradient streams like Tarboo and nearby Chimacum, a dam that blocks the stream's main flow raises the water level upstream and forces it out to the sides—meaning that in high water, when the salmon are running, the floodplain becomes a network of small braided channels. Salmon find these, follow them against the current, and refind the main channel. With that, their end run is complete.

Once the fish have made it to the other side, they're in calm, deep water—a safe place to rest before moving on to navigate the next dam or search for nesting sites.

Beaver dams also do important work when breeding is over and the adults die. When the adult fish are spawned out, beaver dams catch their carcasses, preventing the corpses from washing downstream. They hold the rotting tissue long enough that the nutrients are released and begin cycling through young insects, trees, and fish.

Streams that are ponded with beaver dams also provide the best of all possible worlds for juvenile salmon. Instead of burning up energy swimming against a stiff current, the small fry can congregate in calm, ponded water that is deep, cool, and well oxygenated and where organic material and insect food are abundant. When it's time for the juveniles to make their way to the ocean, they can navigate the dams by wriggling through the leaks.

If you were designing the perfect salmon stream, then, it would start with long stretches of relatively fast-moving water close to the headwaters that provide gravelly spawning habitat. Downstream, there would be even longer stretches of rearing habitat, with a meandering channel broken by a series of beaver ponds.

～

So when that favored cedar fell early in our restoration effort, we didn't respond by killing beavers. Instead, we started to protect selected trees with hoops of chicken wire. As beaver activity in the restoration became more frequent over the next several years, we experimented with different approaches. A long series of failures followed by eventual success told us that the most reliable and affordable system was a 4-foot-high hoop of 1-inch mesh poultry netting held in place with a stout wooden stake. A pattern got established: we protected western redcedars that were close to the creek, thinking that the Sitka spruce needles were too stiff and prickly to be edible. Then the beavers started dropping spruces near the water. So we protected them. Then they took cedars up to 10 yards from the creek. So we protected them. Then they started on spruces farther from the water, so we put wire hoops around them. And so on. They kept us on our toes.

After a couple of years of our gradually making the long-lived conifers we'd planted beaverproof, two small dams appeared. They were just disorganized jumbles of old logs and branches, or what biologists call found wood—not stems that had been cut for the purpose. But still, they were obvious beaver work. And they were holding water.

A week later, Susan and I were clearing a streamside trail so we could walk near the newly ponded area when I found myself sinking rapidly. As I hoisted myself out of the hole, I realized I'd inadvertently collapsed a tunnel leading to a newly excavated bank lodge. We found the second entrance later but were never convinced that the beavers ended up raising a family there.

Two years later, one of the original dams doubled in size almost overnight. We inspected the old lodge again and saw that a small hole had opened in the ground surface. When I stuck my hand down inside it, I found I could sweep my arm around a cavern 3 feet across. Not long after, the dam got bigger and ponded the creek for a good 50 yards upstream. About the same time, beaver-chewed sticks started showing up to cover the hole I'd stuck my arm into. The stick mound grew into a 3-foot-high weatherproof barrier, making it look almost like the lodges you see in the middle of lakes, but on the creek bank. The dam also continued to grow.

The following spring, Susan and I were standing near the lodge, scouting for any recent sign of activity, and heard a series of pips and squeaks. She looked up at the trees, trying to find passing kinglets. There was nothing there, and she looked at me, confused. I pointed down at the lodge—the noises were coming from inside, from beaver kits. We were grandparents in a way. Our boys were relieved when we made the announcement—the pressure was off them for a while.

So beavers are now a feature. The pond they created drowned big trees and shrubs we enjoyed, and even some we planted, but the deaths created snags that attract pileated woodpeckers and red-breasted sapsuckers. The open water has brought in kingfishers, great blue herons, mallards, and wood ducks; it's a paradise for frogs and aquatic insects.

We've also learned to anticipate an intense beaver chew-down sometime each March or early April, right before spring green-up. At this time of year, the parents are hungry from being on short rations all winter and the female is probably pregnant or lactating. Either way, she's eating for a family of four or more. So we have chicken wire at the ready and make the rounds to protect any cedars or spruces we've neglected, trying to direct attacks to willow, salmonberry, and red alder. The willow and salmonberry will resprout efficiently after being cut, and the red alder will seed in without our help, so we consider those species expendable.

When friends and family come to visit now, the first thing they want to see is the beaver dam and pond, and they go hunting for freshly chewed sticks. Everyone wants to see what the beaver family has been up to lately.

~

Beavers were abundant when Europeans landed in North America and took up residence. *Castor canadensis* was found from Mexico to the Canadian Arctic, and from the Atlantic to the Pacific. Extrapolating from their density in the Adirondack Mountains of New York, the naturalist Ernest Thompson Seton estimated the North American population at 100 to 200 million when the *Mayflower* landed in 1620.

At that time, European beavers had been trapped so intensively that they had disappeared from Great Britain. There was still a booming market for their carcasses, though; the fur was being felted to make hats and the scent glands sold to treat headache and fever—they don't help, actually—and as an ingredient in perfumes. The trapping continued, and Eurasian beavers were eliminated from almost all of their former range between 1700 and 1900. By the early part of the twentieth century, only about 1,200 remained. They were threatened with global extinction.

The decline of Eurasian beavers was a boon to the early colonists in North America, though, because beaver hats were still high fashion. The New World species was superabundant, and shipping their pelts to the old country was one of the few ways the early white settlers could earn cash. A cross-Atlantic trade in beaver pelts was under way by 1624.

Beaver felt was used to make the tricorne hats favored by the Colonial army, the top hats worn by aristocrats on outings, and the everyday hats worn by everyone else. By 1800, mountain men were trapping beaver in the Rocky Mountains; soon after, the California fur rush lured the first waves of prospectors to the Pacific coast.

By 1850, the North American beaver was virtually extinct. It was probably raw luck that saved the remaining few: beaver felt went out of style. Silk hats, like Abraham Lincoln's signature stovepipe and the headgear worn by the fashionable gentlemen in Gustave

Caillebotte's famous painting of a rainy Paris streetscape, were in.

The beaver tale is the same as the story of Atlantic salmon and old-growth forests: people exploited a resource until it was gone, or nearly so, and then they moved on to something else.

~

Beavers have come back in North America. The current population is estimated at 10 to 15 million and rising. And the resurgence of beavers isn't isolated. The year I was born, a long list of charismatic megafauna was declining or already extinct in my home state of Wisconsin. The list included sandhill cranes, wild turkeys, fishers (big cousins of mink), cougars, timber wolves, prairie chickens, peregrine falcons, and bald eagles.

In the 1930s, Aldo Leopold wrote the heartrending essay "Marshland Elegy," collected in *A Sand County Almanac*, about the loss of sandhill cranes. His theme was simple: our world is diminished when species like sandhills disappear. But when I was twenty-five, I helped organize a statewide crane count that turned out more than five hundred volunteers—many of them Girl Scouts and Boy Scouts and 4-H kids with their moms and dads. Each team was assigned a wetland that local organizers had mapped, then tasked with arriving at the marsh in the predawn to listen and watch for sandhills. For many of the participants, it meant a 3:00 or 4:00 a.m. wakeup. But that year the census turned up almost three thousand cranes, and five hundred people had a great morning together. Being at a marsh in the predawn of a mid-April Saturday is a memorable experience. The cold is bone chilling, the quiet unsettling, the rising light mysterious. You just sit or stand, watching and listening. Almost always, something happens. It might be a red-tailed hawk floating low over a hayfield or a coyote trotting home after a night's hunt in the woods, or what you came for—the sound of cranes bugling from a hidden recess in the marsh or the sight of a pair flying past, stiff winged, toward their foraging grounds.

Sandhills are now so common that my cousins who farm have trouble keeping them out of their corncribs. And from the steps of Aldo Leopold's Shack today, you can hear breeding pairs of cranes giving territorial calls in wetlands to the north, and hundreds of individuals coming in to roost each evening on the Wisconsin River to the south. In the distance, you can also see and hear a constant stream of east-west traffic on Interstate 90.

Turkeys and fishers were reintroduced to Wisconsin after being wiped out and are now thriving. The Milwaukee-area business leaders who run the Society of Tympanuchus Cupido Pinnatus have purchased enough habitat to give prairie chickens a fighting chance of hanging on in the central part of the state. In 2008 biologists confirmed the first evidence of cougars in a century, and peregrine falcons are nesting on old brewery buildings in downtown Milwaukee.

Almost every state and province has witnessed similar events. When Susan and I moved to Washington in 1985, there were no grizzly bears or timber wolves; both are now being sighted regularly. Comebacks like these are not smooth or easy: in 2009, a father-and-son poaching operation shot and killed the first female wolf to breed in Washington in more than fifty years.

The Tarboo Creek restoration has its own set of comeback stories. We've seen black bear tracks in the mud along the creek and scat in the younger parts of Carl's forest; Peter Bahls took a photo of a bobcat hunting voles near one of our eagle poles; river otter follow the coho run up every fall; cougar are in the neighborhood often enough now to make me nervous walking alone at night; elk were seen just downstream from us recently; fishers have been reintroduced to the Olympic Peninsula and have also been seen in our neighborhood. It's not yet clear, though, whether these animals are coming or going. If the county retains an economy based on timber production and small farming operations, and if citizen action like NWI's plant-a-thon continues, the large mammals and the salmon have a chance. If the area succumbs to the type of exurban and suburban sprawl found in the counties to the south, west, and east, the big animals will be gone.

~

I reread *A Sand County Almanac* every few years and am always struck by the desperate tone of Aldo Leopold's writing. The things he loved and lived for were disappearing, and no one was mourning their loss. He was a prophet crying out in a wilderness. Like Amos or Hosea, he railed at times about the ungodliness of his nation and the retribution to come.

Much has changed in the seventy years since, and much remains the same. In the industrialized countries, most people now profess a strong commitment to environmental protection. There is a growing recognition that the old economy-versus-environment dichotomy is not only tired but also false. Well-organized conservation groups with vocal constituents and access to resources have done a great deal of good. But the deep ethical commitment Carl's father was advocating is still rare. For example, species that have attracted concern, attention, and resources in North America are coming back. But a lot depends on looks. Freshwater mussels have faces that only a mother could love, and they remain in deep trouble. It takes an effective ecological education—and an ethical commitment that borders on the spiritual—to understand that mussels matter, too.

What has changed most since the publication of *A Sand County Almanac* is the globalization of the conservation crisis. Aldo Leopold studied the places he knew best—New Mexico and Wisconsin— and was heartsick at the long-term damage being done to the soils, vegetation, and wildlife. Today, ecologists study habitats around the globe and are overwhelmed by the same emotions. A few popular and well-studied species are being saved, but today's teenagers will probably live to see an extinction crisis parallel to the one triggered by an asteroid that smashed into Earth 65 million years ago.

～

One of the most famous graphs in the scientific field of paleontology has time on its horizontal axis, plotted in increments of a million years. A million years is a long time, but scaled to the 4.6 billion years Earth has existed, it's equivalent to less than a week of your life. This axis starts at about 600 million years ago, when marine animals got big enough to leave fossils that could be studied reliably. (Land plants and fungi appear about 130 million years later.) The vertical axis on this classic graph plots how many species went extinct during each of those million-year intervals. If you look at the trace made by the data, five spikes jump out. These were episodes where more than 60 percent of the species alive at the time went extinct in fewer than a million years. The spikes are called the Big Five and are considered mass extinction events.

The most recent mass extinction was the one that occurred 65 million years ago. Although it was not the most severe cataclysm in history—life forms larger than single cells were almost snuffed out 250 million years ago—the so-called end-Cretaceous or Cretaceous-Paleogene (K-Pg) extinction is easily the most famous. Among other things, it finished off all dinosaurs except the ones that fly—the birds. Although the causes of the other four mass extinctions identified in the fossil record are still hotly debated, the extinction event that closed the Cretaceous period is no longer a whodunit. It wasn't the butler or even Mrs. White in the study with a candlestick. It was an asteroid.

The evidence for an asteroid impact 65 million years ago is overwhelming. We have shell casings near the point of impact—rock fragments that were melted or deformed by the shock waves and heat. We have gunpowder residue in the form of iridium—a mineral that is vanishingly rare on Earth but abundant in asteroids. A dusting of the stuff forms a layer precisely at the point in rock formations where most Cretaceous species disappear. We have bloodstains in the form of soot and ash from wildfires that were triggered by a searing blast of heat following the impact. These are found across the globe in rock layers that are 65 million years old. We have reenactments of the crime—computer models detailing the consequences of spraying enormous volumes of pulverized, sulfur-rich rocks into the air. The sulfurous dust would have triggered acid rain and darkened the sky for months or years, crippling photosynthesis enough to cause mass starvation. Finally, we have the murder weapon itself: a crater that formed when the asteroid plowed into what is now the Gulf of Mexico just north of the Yucatán Peninsula.

A mass extinction like the end-Cretaceous is fundamentally different from what paleontologists call background extinctions. Background extinctions occur at a nonspike rate that varies around a low long-term average. As far as we know, all species eventually go extinct; paleontologists are fond of pointing out that many more species have gone extinct than are alive today.

Although background extinctions are difficult to study, they are routine events. In general they are thought to result from natural selection. For example, one species might die out when mutation and natural selection create traits in a competitor that allow it to harvest resources more efficiently than the declining species can. Or normal rates and types of climate change might eliminate a preferred habitat faster than an unlucky species can adapt to the new habitats that are appearing.

By contrast, mass extinctions appear to occur when the environ-ment changes so profoundly and so quickly—due to some extraordi-nary, not-to-be-repeated circumstance—that evolution can't keep up, meaning that no species has time to evolve adaptations to the novel conditions. Instead, survival is mostly a matter of dumb luck. The birds may have made it through the end-Cretaceous simply because some could fly around to the few habitats that offered anything at all to eat; the mammals that made it may have been able to hibernate through long periods of food shortage. Most other species weren't poorly adapted or somehow biologically inferior—they were simply in the wrong place at the wrong time.

The one pattern that jumps out from research on the Big Five is that species with broad geographic ranges—that are found in a wide array of areas—tend to survive better than close relatives with nar-rower geographic ranges. Apparently, the broadly ranging species are more likely to have individuals living in areas that are not as devastated as most places, meaning that at least some populations get lucky and make it through the environmental crisis.

In the 25 million years following the K-Pg impact, mammals underwent an astonishing diversification that included the origin of the primates. That lineage eventually produced the Old World monkeys, the New World monkeys, and most recently the great apes. The gibbons (also known as lesser apes) and great apes lost tails; the ancestors of today's great apes also gained thumbs that are flexible enough to manipulate objects adroitly. Most great apes also walk on two legs, at least on occasion. The lineage of humanlike species called the hominins walks on two legs mostly or exclusively.

Our own species, *Homo sapiens*, is the only remaining representative of the dozen-or-so species of human or humanlike species in the fossil record. The first of our bipedal ancestors appear in the fossil record about 4.5 million years ago; for most of the intervening period at least two or three hominin species have been walking around Earth at the same time, often in the same place. For most of our lineage's history, it's been unusual to have just one species of human around. Science fictional and medieval-like fantasy worlds populated by elves and orcs and dwarves may have accidentally mimicked reality.

According to the fossil record, all of the human species originated in Africa. But about a million and a half years ago, members of *Homo erectus* emigrated to Asia. They spread throughout the region, leaving stone tools and traces of fire associated with their now-fossilized bones. They stopped leaving fossils, however, about 300,000 years ago. About the time *H. erectus* was disappearing, *Homo neanderthalensis* began to occupy the Middle East, then the Caucasus, and then western Europe. The Neanderthals buried their dead and had larger brains on average than we do; they died out some 40,000 years before the present. The first fossils of *Homo sapiens*, our direct ancestors, are found in African rocks that formed about 190,000 years ago. Early *sapiens* lived among Neanderthals in Europe; they may also have lived just an island or two away from the tiny *Homo floresiensis* in Indonesia—a dwarfed species of hominin nicknamed the Hobbits. At 45,000 years before the present, our *Homo sapiens* ancestors were leaving traces in Australia. And perhaps as long as 30,000 years ago, we'd spread to the New World so that we now occupied every continent except Antarctica. Agriculture developed at several points around the world, independently and with different crops, beginning about 8000 BCE; metalworking in copper started not long after, about 7000 BCE. But the total human population was still small—perhaps 500 million people.

In just the past thousand years, our increased population and ability to alter habitats around the globe has hit Earth like an asteroid.

～

When I teach introductory biology at the University of Washington, we have about fifty minutes to discuss conservation biology. So to give my students a feel for what is happening to biodiversity around the globe, I ask them to apply some of the same mathematical relationships we used the week before to analyze how populations grow through time. For example, if you put $100 in the bank, allow it to earn interest, and find that you have $105 a year later, the annual growth rate of your money is calculated as 105/100, which simplifies to 1.05. The increase is 0.05, or 5 percent, of the original amount. If you left the money alone to continue accumulating interest at the same rate year after year, you could calculate how much money you'd have after a given number of years. The amount of money present at the start of each year gets multiplied by 1.05 to figure out how many dollars you'll have at the end of the year. You do this twenty times to figure out the total after twenty years, fifty times to figure out the total after fifty years, and so on.

Then I ask my students to use the same logic to figure out the growth rate of the number of threatened and endangered species. To do this I show them data on the annual census of threatened and endangered organisms worldwide. The numbers come from the most highly respected agency involved in tracking how species are doing: the International Union for the Conservation of Nature, or IUCN, headquartered in London. The scientists who contribute to IUCN's annual listings use objective and verifiable measures to determine whether a particular species should be considered threatened or endangered. As I write this, IUCN has published fifteen comprehensive assessments over a span of seventeen years. The total number of threatened and endangered species has changed as follows:

1998: 10,553	2006: 16,117	2011: 19,570
2000: 11,046	2007: 16,308	2012: 20,219
2002: 11,167	2008: 16,928	2013: 21,288
2003: 12,259	2009: 17,291	2014: 22,413
2004: 15,042	2010: 18,351	2015: 22,784

To summarize, the number of species that are in trouble grew from 10,553 in 1998 to 22,784 in 2015. Averaged over the seventeen-year period, this is a rate of increase of 1.046—close to the 1.05 annual growth rate of our bank balance, amounting to 5 percent interest. To help my students interpret this number, I point out that IUCN and other authorities estimate that there are 1,560,000 species living today (not including the millions of species in the lineages called bacteria and archaea, simply because they are too poorly studied). If a mass extinction occurred, it would mean that 60 percent of these get wiped out. Sixty percent of 1,560,000 is 936,000.

And then we get to the question of the day: If the rate of 1.046 continues, how long will it take for the number of threatened and

endangered species to reach 936,000 from the starting point of 22,784? When the students do the appropriate plugging and chugging, the answer turns out to be about eighty-three years. I point out that if this is so, it means their grandchildren will live to see the sixth mass extinction in the history of life.

You can argue—strenuously—with the "if this is so" statement. For example, it's almost undoubtedly true that IUCN's estimate of more than twelve thousand species becoming threatened in seventeen years is inflated by what researchers call ascertainment bias. You're probably familiar with this issue; ascertainment bias is responsible for at least part of the recent spike in the percentage of children afflicted with autism spectrum disorders. We're finding more kids with autism, and more species in trouble, because we're looking harder.

Ascertainment bias is important because the eighty-three-years-to-a-mass-extinction conclusion is extremely sensitive to even small changes in the rate of growth of the number of species that are in trouble. You could also object to the assumption that all of the species classified as threatened or endangered, now or in the future, will actually go extinct. After all, we have dozens or hundreds of well-documented comeback stories.

But your analysis shouldn't stop there. For example, I ask my students to think back to the data that they (and you) examined on projections for human population growth, and then I ask whether the pressure on natural areas is likely to be the same in the year 2100 as it was from 1998 to 2015. Most predict that the rate of habitat destruction, and thus the rate of growth of the number of endangered species, is likely to rise. And you could bring up other issues: the total of 22,784 species on the threatened list in 2015 doesn't include the estimated thousand bird species that researchers found recently extinct in Polynesia, or the dodo or auk or passenger pigeon, or the hundreds or thousands of large mammal and bird species—ranging from mastodons and mammoths to Irish elk and moas—that went extinct with our help at the end of the Ice Age. Or that even if the estimate of eighty-three years is off by a factor of 1000, we're still far under the 60-percent-gone-in-a-million-years criterion to qualify as a mass extinction.

Finally, I give my students citations to papers that use other sources of data to address the same question. An array of research teams using different sources of data and a variety of computer models came to the same conclusion independently: the sixth mass extinction in the history of life may be under way.

The key word is *may*. We *may* be on a trajectory to a mass extinction, but that doesn't mean we have to end up there. Beavers have dodged a bullet in North America and in Europe; we've pulled sandhill cranes and grizzly bears and bald eagles and bison back from the brink of extinction. With more restoration efforts like Tarboo Creek and with global action on climate change and human population growth, there's a chance other threatened species have a future as well.

Wild Things

The first full-time job I had out of college, outside of doing carpentry work, was running education and exhibits programs for a little conservation organization called the International Crane Foundation. The group was started by two graduate students who had been studying crane behavior at Cornell University. The two got involved in conservation efforts early in their research careers because seven of the fourteen crane species were endangered at the time. The foundation did field research on four continents; worked to get wetlands set aside as crane refuges in India, Japan, and Korea; and was breeding as many of the species as possible in captivity. They eventually began releasing offspring back to the wild to supplement declining populations or restore lost ones.

It was exciting to be part of an organization in its formative years, though I'll never forget asking the administrator for paper clips and being asked how many I needed so that she could count them out. The most enduring benefit for me was the practice I got teaching—I had to present several hundred public programs each year.

These talks and workshops were an exercise in Aristotle's rhetoric—the art of persuading people to your point of view. I had to convince people that the Crane Foundation was a dynamic, innovative, and effective organization worthy of support. More generally, I had to sell the idea that conservation is good and that it's something people from all walks of life can and should embrace. The tools at my disposal were humor, ribald bird stories (often involving efforts to collect semen from cranes for artificial insemination), and the foundation's long string of successes at protecting wild places.

Of the five or six hundred talks I gave, though, I remember one best. It was a lunchtime presentation to the Rotary Club in Sauk City, Wisconsin. Because my father was a Rotarian, I knew my audience: doers—mostly small business owners—who were committed to their community. So I pulled out all the stops with slides and stories, and ended in what I thought was a blaze of glory. Then came the first question: "Why should we care if cranes go extinct or not?"

It was like throwing a light switch. The room's atmosphere went from bright to dark.

Some of the Rotarians looked upset and embarrassed, but I was sympathetic to the audience member. It's a question I was asking myself almost every morning. At the time, our family's income was so low that we qualified for public assistance. We were eating what my grade school friends called commodity cheese, and we didn't have health insurance. I often asked myself what I was doing.

Aldo Leopold said there are some people who can live without wild things, and some who cannot. I'm not sure that his claim is actually true, given recent research. Experiments have shown that hospital patients heal better when they can see the outdoors, and people who walk in a natural setting have better mental health indicators than similar people who walk in urban settings. An observational study suggests that kids who grow up near green spaces have better cognitive development than kids from comparable backgrounds who don't. But all evidence to the contrary, I have friends, neighbors, and family who are convinced they can live without wild things, and I suspect that almost nothing could persuade them otherwise. If the man who asked the question was like them, there probably wasn't much I could do or say. But I had to try. So I acknowledged the issue as an important one and said I'd come up with three reasons.

The first was practical. The Crane Foundation was really in the business of preserving wetland ecosystems—the birds were just a way to get peoples' attention. In addition to hosting wildlife, intact wetlands minimize flood damage, regenerate groundwater, and purify surface water. They perform important public services and save us money. If the United States had been smart about protecting wetlands as its cities developed, we'd be saving hundreds of lives and many billions of dollars in flood insurance and flood-induced repair bills every year.

My second reason was ethical. The people I admire most take responsibility in their work, and in their communities and families. They make things better by giving something back to the world. They also live a moral life. The essence of morality is to care for things other than yourself, treat other people and living things with respect, and use power wisely when you have it. Biological and cultural evolution has now put human beings in a position of immense power relative to other species, and we can be destroyers or stewards. This is a moral choice. I view practicing conservation as a matter of goodness, of doing the right thing. It is wrong to harm a species or an ecosystem. If we are the reason cranes and other species are going extinct, we need to take responsibility and act on their behalf.

The third and final reason was more personal. It was aesthetic, even spiritual. People enjoy living in places like Sauk City because they are beautiful. The town is surrounded by productive, well-managed, family-owned farmland and rolling woodlands that are large and intact enough to support important populations of rare birds and mammals. Dozens of bald eagles spend the winter along the Wisconsin River just north of town. I asked the group to imagine how poor their lives would be without these things—if they lived in a world of factory farms and asphalt. Fundamentally, I said, we work to save cranes and wetlands for the same reason we work to help our communities, friends, spouses, and children: because we love them.

~

More than thirty years have gone by since that afternoon in Sauk City. I'll never know whether I made an impression, positive or negative, on the questioner. But during this span, ecologists have done hundreds of experiments to test my first claim—that conservation has practical benefits. The initial question was whether natural communities with many different species function better than the same areas with fewer species. The motivation was simple: researchers realized that we are losing many organisms to extinction and wanted to know if it matters biologically.

The first papers summarized experiments in prairies. The basic scheme was to set up an extensive array of experimental plots in the same area, sow different numbers of grasses and other flowering

plants in each, let everything grow for a couple of years, and then measure productivity—the grams of biomass, or green stuff. For practical reasons the researchers had to ignore what was going on with root growth and limit themselves to measuring aboveground biomass. What they found, in experiment after experiment, is that productivity increased with the number of species present in each study plot until a saturation point was reached. The result was important because plant biomass is what everyone else, from fungi to bison, eats. In general, productive plant communities support a larger number and more diverse array of animals.

These studies were quantifying what biologists and economists now call ecosystem services—the "goods" natural areas deliver to humans and other species. In the 1930s, Aldo Leopold wrote about the same thing but called it land health. The problem that he and other ecologists had at the time was quantifying what they meant by health—coming up with objective, rigorous ways of testing the hypothesis that species-rich areas work better than species-poor ones. Although the soil erosion problems of the Dust Bowl years were starting to open some eyes, at that time almost no scientific evidence existed on the question of whether species diversity matters. Rather, it was an article of faith that leaving some areas undisturbed would provide tangible benefits for humans. And in general, the idea of managing land for something other than the maximum and most immediate economic gain was new, revolutionary, and largely untested.

More recent experiments have documented that when a drought or other disturbance occurs, productivity declines less in species-rich habitats than in species-poor areas. Places with high species diversity also tend to recover from disturbance faster, resist invasion by exotic species better, and sop up more carbon dioxide from the atmosphere. The broad implication here is that a mass extinction will make Earth less stable and productive.

It's unlikely, though, that this research will inspire people to do the things it will take to preserve biodiversity: limit family size, favor locally grown food, prefer smaller and more efficient homes, transfer the time and money spent on driving to more enjoyable activities, and contribute to the preservation of wild places through taxes and private donations.

Scientists are trained to be persuaded by data, but for most people it's not about the evidence. It's about values: what we want and what we think is good.

～

The why-care question from Sauk City will get more common as today's generation grows up. Since the late 1980s there has been a slow and steady decline in the probability that an average American will visit a national park, state park, or national forest. The same trend occurs in the proportion of people buying a fishing license or duck stamp. The number of hunters is holding steady—perhaps because deer populations have exploded in many parts of the United States. The same downward trend in national park visitation is occurring in Japan, the only other major industrialized country for which data are available. Both of these analyses have controlled for changes in population size: even though the total number of visitors may be up in some areas, the overall percentage of people who seek out direct experiences with nature is declining.

If people care about things they love, and if a love of the outdoors grows from contact and experience, a smaller and smaller subset of the overall populace can be expected to extend themselves to save patches of the natural world.

This realization hit home a few years ago after a series of planning meetings for the Seattle Parks Department. I was part of a citizens' group that was advising on the design of a new park being created from a recently decommissioned U.S. Navy air station. The Parks Department and the designers were proposing an intensive-use area with lighted playing fields next to a low-impact area with what would become one of the largest wetland restorations in western Washington. Our boys were playing soccer and baseball at the time, so my official capacity at the design meetings was to represent the needs of youth sports organizations. But I was thrilled with the wetlands proposal as well—it would be a great place for people to stroll and bird-watch, and wonderful shallow-water habitat next to the deeper-water environments of nearby Lake Washington.

At one particular meeting, the people advocating for the wetlands had been rhapsodizing about the amount of frog habitat that would be created. The other ball fields person there was nonplussed. "What," he asked as we left, "is the big deal about frogs?"

I thought of him recently when I opened the door of our toolshed at Tarboo Creek and grabbed a box of galvanized nails from a shelf. Crouching in the corner, no longer in shadow, was a northern Pacific tree frog—a black-masked creature about an inch and a half long. You already know its call; it gives the classic "ribbit" that movie producers use in soundtracks for night scenes. The little frog in the shed was khaki colored, blending in with the wood shelf and walls. But out in the restoration area, not a stone's throw away, I'd just found a grass-green tree frog of about the same size. Tree frogs can change color to match their background. They're not as fast as a chameleon, but they will change before your eyes if you are patient enough to sit and watch for ten to fifteen minutes. I love the 'tweener stage best, when they're not brown and not green but a mottled neither-nor.

We hear tree frogs almost every month of the year at Tarboo. On cold autumn or even winter nights, the odd call will sound out of nowhere, from somewhere high in a bigleaf maple tree. We remark on the one brave soul out in the darkness, body chilled to air temperature but still managing one unanswered call before quieting for the night. Then in spring, we use the volume of the frog chorus to track the temperature in the two ponds on the hill above the creek. Susan and I both grew up in the Midwest, so we experienced spring as a two-week orgy of budding, buzzing, and bursting before the forest canopy snapped shut with leaf-out and summer set in. But spring in the Pacific Northwest is a three-month marathon, with the temperature inching along day by day and week by week: two degrees forward and one degree back. The frog chorus is an audio thermometer, letting you know how things are going. On cool nights the calls are sporadic, as if cued by a conductor beating time and then pointing at random intervals around the orchestra. When warmth comes, though, everyone sings all at once and as fast as possible, so the calls swell and blend into one throbbing nightlong aria. We tiptoe along the path next to the ponds, arm in arm, trying to get close enough to pick out individual calls near our feet. Invariably, we do something that shuts down the chorus, bringing sudden silence. We never know what it is; sometimes the chorus quiets as we stand stock still, barely breathing. It can be many minutes before one courageous, love-struck male breaks the stillness. Then others join in—sporadically at first but with voices growing in frequency and volume as confidence builds. At last the assembly relaxes and throws itself back into the business of calling and mating.

And then there are tads. We begin looking for egg masses in April—gelatinous blobs that cling to sticks and leaves in the ponds' shallows. The black specks inside are eggs. At hatching, the froglets work their way out of the jelly into the open water and begin feeding on bits of organic debris, microscopic crustaceans, and tiny algae. When they metamorphose—losing their tails, growing fore and hind limbs, and remodeling their gills and lungs and skin to make air-breathing possible—they are still tiny, perhaps half the size of a dime. At this stage they disappear into the wet grasses near the ponds and are lost to us for weeks. But sometimes in late summer they reappear in numbers. We never know why, but some years are froggy; the shores of the drying ponds are alive with thousands of little tree frogs, green and black and brown. It can be hard to walk for stepping on them. We sometimes just sit there, marveling and watching.

To understand what the big deal about frogs is, then, you have to develop a connection—a sense of the ecological and spiritual bond among everything that is or has been alive.

Snakes, since the time of Eve, have been a particularly hard sell for people. Alexander Skutch, considered one of the first great tropical ornithologists, killed them on sight. But even people with debilitating snake phobias can be treated—by exposure.

Western Washington is not snaky—the cold and damp make fungi, mosses, and spiders the local celebrities when it comes to biodiversity—but garter snakes are common. Our older son is expert at catching them and has been since he was three. He quickly learned to get a second hand on them and hold the wriggling rope away from his body, as garter snakes spew a noxious blend of feces and slime when they're grabbed by a predator.

One afternoon several years ago we were out walking the pond's edge checking for tree frog babies and other aquatic life, not actively looking for snakes. A couple dragonflies were whizzing about; some water boatmen were bobbing up and down in the shallows; and then, apropos of nothing, a snake's head popped up in the middle of the pond and began periscoping around. Off and on I could see the rest of the body undulating in the water. Then the snake expertly plunged back below the surface. Seconds later it reappeared, a yard or two off. It was hunting tadpoles. I'd never seen the behavior before.

I've had other close encounters with garter snakes. When I was in college I went running through the arboretum adjacent to campus on an early spring afternoon and surprised a red-tailed hawk that was standing on the ground. It had been thrashing about but stopped to look at me. Its beak was bloody, and it quickly flew up and away before I got closer. When I went up to where it had been standing, I found two large slabs of limestone with a shallow slit under them lined with old leaves and writhing with hundreds of garter snakes. It was a hibernaculum, where snakes gather to sleep away the winter. Apparently the hawk had discovered the cache as the first few chilly, drowsy snakes were beginning to emerge. Now the ground around the mini-cave was strewn with bloodied pieces of snake. The hawk had gone into a killing frenzy—stomping at the snakes with its talons and slashing at them with its beak.

At Tarboo I once killed a garter snake accidentally by driving a truck over it. When I picked up the body to examine it, I noticed a head sticking out of the cloaca—the single opening near the end of the tail. By running my thumb down the body, I was able to deliver seven baby garter snakes, unfortunately all stillborn. Among snakes and lizards, it's common for populations in northern or high-elevation habitats to give birth to live young rather than laying eggs. The most compelling explanation is that in cold environments, females do better by producing smaller numbers of larger young, protected inside their bodies, versus a larger number of eggs that have to brave the cold on their own. It's a quality-versus-quantity trade-off. And if the growing season is short, better-developed young have an increased chance of getting big enough in a summer to make it through their first winter. In general, larger animals are also better able to cope with cold temperatures than smaller individuals—they have a heat-retaining volume that is large relative to the heat-losing surface area. I'd known these things for years but hadn't realized that garter snakes give live birth until I saw the young emerging from that dead mother's body.

And I will never forget the time a ten-week-old black lab puppy— one we were raising for a service dog agency—discovered a snake. It was early March, and the puppy found the first snake I'd seen that year. The snake was moving slowly on the cold ground; otherwise it would have disappeared at the first sight of a dog, even such a little one. As it slithered away with agonizing slowness, the puppy sniffed and pawed at the tip of the snake's tail, trying to figure out what it was. The puppy was both fascinated and frightened, cautiously lowering its head to give a tentative sniff and then jerking backward at the first twitch. I stood by, smiling. Puppies and small children help you see the world with fresh eyes.

Exposure, along with an open, inquiring mind, is the start of understanding the bonds among tree frogs, garter snakes, and people. The key is to understand those bonds well enough to feel their presence.

～

One of the great values of working on Tarboo Creek is exposure. If you're out planting or removing invasives or pruning saplings, you notice things—or things happen to you. I was sitting on a stump once, taking a break from thinning trees, and realized that three black-tailed deer were standing 10 yards away, staring at me with bulging eyes. I stared back. There were two does and a young buck, its little rack of antlers in summer velvet. The three sniffed and peered even more intently, trying to decide if I was dangerous. But the situation was difficult for them, as I was downwind. Finally, the closest doe decided I was indeed one of the two-leggeds and ran off. Actually, she stotted away—in the stiff-legged, bounding, thumping jumps made famous by the Thomson's gazelles of East Africa. Tommies stot when they see a predator approaching, long before the cat or hunting dog is actually within striking range. It's a signal thought to communicate "I've seen you and am well away—don't waste your time, and my energy, giving chase." White-tailed deer will do an analogous display, lifting a stumpy white tail and waving it like a flag, taunting you as they saunter away. I've also seen white-tailed deer tuck their tails flat against their thighs when I've surprised them at close range and they're running for their lives. But I'd never before seen a deer stot. I felt privileged.

These exposures remind me of one of the fundamental messages of ecological research: all organisms are connected. Sometimes the connections are intimate, as with the human hunters who feed their families venison sausage, or the fungi that wrap themselves around spruce roots, or the bacteria that live inside alder nodules. And sometimes the connections are as direct as the food chain that links water bugs with tree frogs with garter snakes with hawks. But even if the connections are loose, they are present. We buried a family dog under a cedar tree at Tarboo Creek. By now, most of the nitrogen atoms in her body have been transported up the trunk to help make new leaves, where they lived for a year or two before dropping back to the soil. Someday soon those same atoms may wash down Tarboo Creek to Puget Sound; decades hence a salmon may swim them up a watershed in northeast Siberia. The idea that any organism lives and acts independently of others is a myth. The realization that all organisms are connected is a profound insight.

~

Ecologists study both the biotic connections among species and the physical connections between organisms and the abiotic environment—the air, soil, weather, and water. But species are connected by common descent as well. This is a fundamental message from research in evolutionary biology. To the best of our knowledge, all organisms living today are descended from the same common ancestor. There is no such thing as an unrelated species; we are all blood relatives.

The proteins that repair damaged DNA, and that can predispose people to skin cancer when they're not working properly, were initially discovered and studied in the bacterium *Escherichia coli*. The proteins that regulate cell division, and that predispose people to cancer when they're not working properly, were initially discovered and studied in the yeast *Saccharomyces cerevisiae*. The similarities aren't accidental—they're the products of shared evolutionary history. *E. coli*, *S. cerevisiae*, and humans all share a common ancestor that had these DNA repair and cell division proteins.

The realization that all organisms are related is fundamental to understanding our world. But there is a second, equally important, result from evolutionary biology. For thousands of years, perhaps beginning with Aristotle, philosophers and scientists had imagined that organisms exist in a hierarchy—a ladder of life, also called the Great Chain of Being. The idea was that some species were lower because they were judged to be simpler or less developed. Even after the advent of evolutionary biology, people clung to the idea that there are higher and lower organisms, even using sloppy language like "less evolved." In the late 1800s this kind of thinking was used to justify

colonialism and slavery, because it was assumed that well-educated people of western European ancestry were at the top of the ladder.

Evolutionary biologists have shown that this type of thinking is incorrect. There is no such thing as a higher species or a lower species, because different types of organisms aren't arranged as rungs on a ladder. Instead, species are twigs on a tree of life. Humans aren't at the top of a ladder, we're just a twig among twigs—one species among many millions. To drive this point home, consider what you'd discover if you traced the past few generations of your family tree back further and further and further. Eventually you'd connect to the common ancestor of all humans living today. Keep going and you'd connect that ancestor to its ancestors. At the base of this tree, all organisms trace their ancestry back to the same initial life form. Researchers call this organism LUCA, for last universal common ancestor. If you look at people and at the *E. coli* living in the guts of people, and if you trace the ancestry of each species back to LUCA, you'll realize that the same amount of time has passed—both have been evolving for billions of years. They are not higher or lower than one another; they are simply adapted to different environments and ways of life. So evolutionarily, there's nothing special about human beings. We're just a species among species.

And as for you and me? We are links in a chain: part of a single, enormous family tree that dates back billions of years and includes all life forms, living or extinct.

Ecology tells us we are part of a community; evolution tells us we are part of a family. Earth is the only place in the universe where life exists, as far as we know, and all of that life is unified: through our interactions in a common environment and through sharing a common history. We are not separate or isolated—we are part of a whole. Listening to tree frogs on a warm spring night or seeing a black-tailed deer stot through a forest affirms that unity. It also reminds me that humans and other living things are connected emotionally as well as ecologically and evolutionarily.

~

There is a parallel between what ecological and evolutionary research has revealed about the nature of life and what the events of the last hundred years have revealed to people. After thousands of years of conquest and exploitation, of emphasizing the differences among nations and races and ethnic or cultural groups, the people of the world have never been so closely connected. Economically and politically, we are more intertwined than at any other time in history. Our world has shrunk; today no woman, man, community, or nation is an island.

We are also entwined genetically. Research since the mid-1990s has shown that the genetic or biological differences among human races are trivial compared with what is routinely observed among populations of other animals and plants. To understand why, you have to look at data from the fossil record and genetic analyses. Both sources of evidence agree that humans originated in Africa, then colonized the Middle East and then either the Caucasus and western Europe or, in a separate wave of emigration, southern and southeastern Asia, then northern Asia and the New World. In effect, a subset of humans left Africa, and subsets of that subset broke off to head north or east. Along each major corridor of human expansion, some individuals stayed and some moved on to new areas over time. Each group that emigrated was a small genetic sample from the group that stayed, and an even smaller sample from the group that originally left Africa. As a result, there is still more overall genetic diversity among people in Africa today than there is in the rest of the world combined. As researcher Svante Pääbo points out, genetically we are all African.

It is true that novel genetic differences arose as humans evolved over time, in Africa and elsewhere. Worldwide, the genes involved in hair coloration have versions—or alleles—that code for blacker or browner or redder or blonder hair. But the handful of alleles responsible for the differences among us in facial features, hair color, and skin color are dwarfed by the tens of thousands of alleles that we share by common ancestry.

Genetic differences among the human populations that scattered over the globe were small to begin with and are beginning to disappear as we emigrate, immigrate, and intermarry. The degree of genetic distinctiveness among human populations probably peaked several hundred years ago and is being erased by a biological fusion.

As the fall of the Iron Curtain and other political movements indicate, an economic and cultural fusion may also be in its very early stages—one based on free trade, democratic governance, a commitment to education, embracing differences in religious belief and cultural practice, and equal rights for women and ethnic minorities. People who benefited from the old order are fighting this change bitterly and often violently. They will continue to fight. But it is possible that an enduring bond can be built among peoples and nations, and it is possible that it will be based in part on the tasks we do together to preserve biodiversity and mitigate climate change. Stewardship and restoration—a commitment to the value of wild things—could be part of a great common work that reinforces the evolutionary and spiritual bonds among peoples, and the ecological and economic bonds among nations.

It is possible. But is it probable?

A Natural Life

Newscaster and author Tom Brokaw called Carl Leopold and his peers the Greatest Generation, with good reason. They survived the Great Depression, fought a world war that defeated fascism, sponsored the Marshall Plan that laid the groundwork for a peaceful and united Europe, funded the largest expansion of public education and scientific research in history, endured 44½ years of a Cold War that eventually brought down communism and freed Eastern Europe, unleashed the power of nonviolent protest to end legalized racism, and built the most prosperous economy the world has ever seen.

The values that Carl's friends fought and sometimes died for in Normandy and Guadalcanal were simple: dignity and self-determination. They fought against tyranny and for liberty. But after enduring the privations of a depression and the horror of a world war, the Greatest Generation reacted by pursuing a different value: comfort. After sacrificing their youth and risking their lives for others, they wanted to focus on their own well-being. The American Dream morphed from its original conception—which historian James Truslow Adams had articulated as a quest for personhood, self-fulfillment, and citizenship—to consumerism. This new and extreme emphasis on the individual and the material led their children, the Baby Boomers, into gated communities where three-car garages were required by zoning regulations. Writer Tom Wolfe called the children of the late 1940s and '50s the Me Generation, with good reason.

This new version of the American Dream was exported. As the economies of India and China boomed in the early 2000s, their expanding middle and upper classes wanted to live like Americans—the ones in the gated communities. Worldwide, the good life wasn't being defined by goodness. It was being defined by consumption.

~

One way to diagnose what's important to a community is to look at how it spends its money. The most recent analysis of spending for land conservation in the United States showed that in 2001, federal and state government agencies spent about $2.74 billion on the permanent protection of land. Conservation biologists estimate that this total would need to at least double to create an effective network of strategically preserved areas, meaning the lands that give the highest return on investment in terms of preserving biodiversity.

To put these numbers in perspective, Americans spend more than $6 billion per year, or about $70 per person, on Halloween costumes, cards, decorations, and candy. And Halloween represents the smallest expenditure on any major celebration; total spending on holidays is $228 billion per year. Although data on sales of pornography are notoriously difficult to collect, a 2001 study by *Forbes* magazine gave a conservative estimate of annual sales at about $2.9 billion. If so, Americans spent more on pornography that year than on land conservation purchases and easements.

~

Something about the way human beings are wired makes us love dichotomies. We are asked to choose between economic growth and environmental quality. Depending on the region involved, politics are defined by hard-edged extremists on the left and the right or marketed as a stark choice between the security promised by despots and the chaos and sectarianism of emerging democracies. In pop psychology, personalities are type A or type B.

In my own field of evolutionary biology, a dichotomous nature-versus-nurture debate raged for more than fifty years. The question was whether the variation we observe in human behavior is due to variation in our genetic makeup or to variation in learning and other aspects of the environment that we experience. The issue spluttered out when researchers finally realized they were arguing about a nonquestion. In the history of life, no gene has ever been expressed in the absence of environmental influences. These environmental influences, in turn, come in two distinct but interacting types: a genetic environment created by the unique combination of fifty thousand distinct alleles in every person, and an external environment with thousands of factors—temperature, disease, nutrition, learning experiences, social milieu, and so on—that influence whether a gene is expressed at all and if so, when, where, and how much. Likewise, it's not possible for experience, learning, training, or other aspects of nurture to create a trait in the absence of the structures and potentials created by gene products. There is simply no such thing as nature *versus* nurture.

Most dichotomies are just as sterile and misleading. So even though we adore our *or*s, reality is really about *and*s and *both*s and degrees. We need to explore positions on a continuum and seek balance. Aristotle understood this when he called the golden mean golden.

It would be naive and counterproductive, for example, to claim there is a stark choice between experiencing economic well-being and enjoying a healthy environment. No one wants to promote poverty, and no one wants to live in a place where soils, plants, and wildlife have been destroyed. A quick glance around the world should convince you that poverty thrives in places where the soils, native plants, and wildlife have been obliterated.

Still, it's not difficult to find evidence that something is badly out of balance in the current version of the good life—the one based on consumption.

According to the U.S. Centers for Disease Control (CDC), more than 11 percent of Americans age twelve and up are taking prescription medications for depression—a serious, sometimes life-threatening, illness. Physicians recommend these drugs in response to combinations of the following symptoms:

- persistent sad, anxious, or empty feelings
- feelings of hopelessness or pessimism
- feelings of guilt, worthlessness, or helplessness
- irritability or restlessness
- loss of interest in activities or hobbies that are usually pleasurable, including sex
- fatigue or loss of energy
- insomnia or excessive sleep
- overeating or appetite loss
- thoughts of suicide
- difficulty with concentration, remembering details, or making decisions

The CDC's 11 percent estimate doesn't include people who are self-medicating for depression with alcohol or street drugs. The U.S. National Council on Alcoholism and Drug Dependence (NCADD) reports that about 20 percent of people age twelve and older use prescription painkillers, sedatives, and stimulants for nonmedical purposes; in addition, more than 8 percent of American adults abuse alcohol, and 8 percent of individuals over the age of twelve use street drugs. If you live in the United States, you are almost undoubtedly in daily contact with dozens of people who are drug dependent.

In addition, a small but growing body of research suggests that some people are using food to self-medicate for depression and anxiety. Although studies on the extent of food addiction and its causes are just beginning, early data from Europe and the United States indicate that symptoms of it occur in 10 percent of the general population and almost 40 percent of obese individuals. The global obesity epidemic itself is well documented: the most recent data indicate that worldwide, rates of overweight condition and obesity have increased by 28 percent in adults and almost 50 percent in children since 1978. In the United States, about one-third of the adult population is obese; rates in China and India are not far behind.

Obesity is complex, but at a fundamental level it is a disease of excess. Depression and depression-induced alcohol and drug abuse are also complex diseases, but they don't simply reflect economic poverty; the rates reported by the CDC and NCADD are from the wealthiest society the world has ever seen. Instead, we are suffering from diseases that are rooted, at least in part, in a poverty of values. We have conflated being well-off with well-being.

～

People study happiness. To date, one of the most robust conclusions from this research is that income matters—but only up to a point, and a relatively modest one at that. In the United States, measures of emotional health improve as annual family incomes rise to the $50,000-to-$75,000 range, then level off. Given the messages we are bombarded with in electronic and print advertising, this result is surprising. The message from ads and pop culture is that wealth is the key to happiness and that extreme wealth is the ultimate source of good feeling and self-worth. But the message in the data—in what people actually experience, instead of what advertisers want us to think—is that happiness starts with having your basic needs met and achieving a moderate sense of financial security and independence. After that, pursuing intrinsic goals like spiritual growth, healthy relationships with family and friends, and rewarding work—in contrast to extrinsic goals like fame and fortune—is what creates happiness, meaning, and fulfillment.

Commentator Arthur Brooks points out that in terms of pursuing happiness, we are wired for the wrong things. Biologically and psychologically, our predisposition is to love things and use people. Reviewing data on deforestation, overfishing, habitat loss, and extinctions, I would rephrase his summary slightly: we're predisposed to love things, and use people and natural resources. But research in sociology, psychology, and economics shows that to actually live well we should flip the order of our proclivities: we should love people and nature, and use things. Used in moderation, things are a means; in excess, they are an end.

In terms of being based on rigorous research, these conclusions are new. But they are also ancient. They are the guiding principles of the Buddhist, Christian, Hindu, Jewish, Native American, and other enduring spiritual traditions, as well as the foundation of Aldo Leopold's land ethic. Jesus Christ was particularly explicit. It is harder, he said, for a rich person to achieve humanity's true goal in life—entering the Kingdom of God—than it is for a camel to pass through the eye of a needle. When he preached on this point, he repeated it for emphasis. It was the only time he did so, according to the teachings recorded in the Gospels.

Every generation has to find its own way of fighting our materialistic nature and reminding its children of the values that matter and endure. We are born to take, but we learn to give.

So we have work to do.

If you are entering adulthood right now, you will need to find a balance between the consumer life that dominates global media and advertising and a more natural life. To do so, you will need help from elders—parents, grandparents, teachers, and mentors. A more natural life is a braver choice, because it doesn't conform to the values that pervade popular culture. In that sense, Thoreau was correct in saying that people as a whole don't yet lead a natural life. But it is more than possible to start, by joining the thinking community working toward that goal. One of your first steps could be to surround yourself with people who organize their lives around four beatitudes.

BE ENGAGED. A natural life includes work that connects you to the land and to friends and family who care about the land. That work may be hunting or fishing for some of the food you eat, or gathering or growing it, or doing citizen science, or planting trees or saguaro cactus or bottle gentians, or giving time and money to organizations that do good work for people and the land.

A natural life is active. It is one in which you invest time, energy, skills, and money on behalf of people in need and the conservation of living things. This is our primary focus at Tarboo Creek. It's our place to engage in work that can help heal a damaged landscape—a place for us to give back and, by so doing, increase in wisdom, happiness, and understanding. The work works at two levels: saving salmon and trees, and sustaining personal growth.

A natural life isn't something you talk about. It's something you do. If you choose the path that leads to a more natural life, one of your first steps may be to find your own Tarboo Creek.

BE SIMPLE. It takes self-confidence to choose a natural life, because you have to define yourself by who you are instead of what you have. You may have to resist what advertisers want you to want, and avoid comparing yourself to others who have bigger, newer,

or more things than you do. One of the great ironies of having this self-confidence, though, is that you will probably live better by living more simply. It's common, for example, for people to put a large dream home at the top of their list of life priorities. An acquaintance recently bought his; to afford the required square footage, he had to move far from his former home. He now drives ninety minutes to work each way instead of the thirty minutes he used to walk and bus from an apartment. The house is costing him two hours of commuting every day, or ten hours per week. Each year he is giving up five hundred hours that he could be spending with his family. It's as though he were offered twelve and a half weeks of extra vacation per year, every year of his working life, but turned it down.

There are a million other ways to practice simplicity besides living in a comfortably sized space. Aldo Leopold was famous for pulling most of the building materials for his family's Shack out of the Wisconsin River—rescuing orphan boards that floated down and stuck in sandbars. He even wanted to name his book *Great Possessions*, because its core message was about the richness of living a life focused on your relationships with people and the land instead of trying to create meaning by acquiring things. As his daughter Nina used to say, "The Shack was nothing, but it was everything." It wasn't a life of denial. It was a life of fulfillment. The family possessed great happiness there.

Being simple is a commitment to reducing the importance of the material, physical parts of your life and increasing the importance of your friends, family, community, and spiritual growth.

BE REAL. The opposite of a natural life is an artificial life—one that is marked by posing. An example parks outside my home every morning in the form of a large sports-utility vehicle that an employee at a nearby store uses for commuting. The advertising campaigns for this SUV picture it in mountainous terrain or towing hunting boats. It costs $50,000 new; a typical five-year loan pushes the price to $55,000, including interest. A natural life avoids this type of spending, which economist Thorstein Veblen called conspicuous consumption. If you actually are a rugged outdoorsperson who climbs mountains and hunts or restores habitat, you don't need a vehicle like this to pose as one. If you are true to yourself, you don't need to spend money on highly visible badges of status that are meant to let everyone know how wealthy you are or who you wish you were. Those types of badges simply aren't important to you.

Living a natural life should provide the peace of mind that comes with being yourself instead of adopting an advertiser's image of what you should be. I remember Carl Leopold praising a friend by saying that she was "comfortable with herself." The quiet self-assurance he was commenting on is a quality of people who understand and accept themselves. But this type of humble self-assurance can be rare. In 2014, the American Society for Aesthetic Plastic Surgery reported that Americans spent more than $12 billion to get their breasts, faces, or waists altered. The ten most popular interventions alone totaled more than 8.1 million procedures, all with the goal of helping people pose as someone more beautiful—usually as defined by fashion models or celebrities.

A natural life begins with loving yourself for who you are. Posing has psychological, physical, and spiritual costs that you avoid in a natural life. Being simple is a commitment to using fewer resources; being real is a commitment to self-affirmation and self-acceptance.

BE PRESENT. Being present in the here and now is a major goal of prayer, meditation, worship, yoga, and other forms of spiritual practice. Religious people have strived to immerse themselves in the present for centuries.

Becoming alert to the present has always taken focus and effort. But it is even more difficult to achieve now, because of the distractions offered by our electronic devices. I was working in our garden recently and noticed a small procession coming my way. It was led by a little boy wearing a brand-new bicycle helmet and riding a tiny two-wheeler with training wheels. A gray-muzzled golden retriever trotted by his side; a dad brought up the rear and held the dog's leash. It was a beautiful, sun-splashed spring day, with crocus peaking and daffodils opening for the first time. I looked up to say hello but couldn't catch an eye—the father was staring at a phone he was holding about 8 inches from his face. He would walk a few paces, stop to read a message, walk a few more steps, stop, and scroll to the next message. The boy and the dog had to start and stop with him, over and over. I watched as the little family went around the corner and disappeared, the father still stopping and starting, transfixed by his phone.

You can see a similar phenomenon at the Museum of Modern Art in New York City every Friday night, when admission is free. Hundreds of people pack into the galleries, surrounded by some of the greatest paintings ever made. But almost no one is looking at them. Instead, all but a handful of people spend their visit taking stills or videos of the art. If there is a particularly famous work like Van Gogh's *Starry Night* or Picasso's *Three Musicians* in the room, one person after another will have their photo taken standing next to it, then move on.

When the Leopold family was working at the Shack in the 1930s and '40s, they practiced being present by recording phenology, or the timing of natural events. They would document the first goose flocks to arrive in spring, the first *Baptisia* flowers to open in summer. Nina Leopold made parallel observations when she retired to the property thirty years later, and she subsequently co-authored a scientific paper showing that many events at the Shack property, including dates of first flowering and dates of arrival for migrant birds, were speeding up— presumably in response to global warming.

Aldo Leopold was also famous for gently challenging himself and students or friends or family members to look hard at what was around them when they were out walking, and to read

the landscape. "Who do you think made this track?" "What could've made this tunnel through the grass?" "Is this deer scat new or old?" At Tarboo Creek, our younger son practices being present by taking a short walk in the restoration area every evening, by himself and without his phone, before he goes to bed. Even in the heart of the city, you can be aware of cloud formations, the phases of the moon, the flow of a local river.

If you are present, you can and will notice things—about the world and about yourself.

～

The good life, as a natural life, is ecological in outlook and practice. The overriding principle is to be part of a community—to belong to something larger than yourself and participate in it. A natural life focuses on minimizing how much you take from the land and other people, and maximizing how much you give. It is a life of humility.

To prevent a mass extinction and widespread human suffering, the generation that is entering adult life now will have to be the greatest generation of all. The path we have followed for centuries has led to resource depletion, deterioration of soils and water, climate change, and extinctions triggered by human overpopulation. If you are beginning your adult life, or mentoring a young person starting theirs, this book is a call: please, help us find a new way.

Acknowledgments

~~~~~~~~~~~~~~~

I treasure the advice and encouragement I received from readers of early versions of the manuscript: Owen Fairbanks, Ben Freeman, Peter Freeman, Teri Hein, Allan Kollar, Konrad Liegel, Jim Smith, Sarah Spaeth, and especially Lorraine Anderson, Deborah Easter, Linda Gunnarson, and Mary Kollar. Susan Leopold Freeman has been my partner in life and in work—almost everything we do at Tarboo Creek is done side by side.

References

INTRODUCTION: NOTICING THINGS

Christie, P. J., D. J. Mennill, and L. M. Ratcliffe. 2004. Chickadee song structure is individually distinctive over long broadcast distances. *Behaviour* 141: 101–124.

Diamond, J. 2005. *Collapse: How Societies Choose to Fail or Succeed.* New York: Penguin Group.

Goodale, E., and S. W. Kotagama. 2008. Response to conspecific and heterospecific alarm calls in mixed-species bird flocks of a Sri Lankan rainforest. *Behavioral Ecology* 19: 887–894.

Harrison, N. M., and M. J. Whitehouse. 2011. Mixed-species flocks: An example of niche construction? *Animal Behaviour* 81: 675–682.

Magrath, R. D., B. J. Pitcher, and J. L. Gardner. 2007. A mutual understanding? Interspecific responses by birds to each other's aerial alarm calls. *Behavioral Ecology* 18: 944–951.

Soard, C. M., and G. Ritchison. 2009. "Chick-a-dee" calls of Carolina chickadees convey information about degree of threat posed by avian predators. *Animal Behaviour* 78: 1447–1453.

Sridhar, H., G. Beauchamp, and K. Shanker. 2009. Why do birds participate in mixed-species foraging flocks? A large-scale synthesis. *Animal Behaviour* 78: 337–347.

Templeton, C. N., and E. Greene. 2007. Nuthatches eavesdrop on variation in heterospecific chickadee mobbing alarm calls. *Proceedings of the National Academy of Sciences* 104: 5479–5482.

Templeton, C. N., E. Greene, and K. Davis. 2005. Allometry of alarm calls: Black-capped chickadees encode information about predator size. *Science* 308: 1934–1937.

A STREAM IS BORN

Ambrose, S. E. 1996. *Undaunted Courage*. New York: Touchstone Books. (For information on Lewis's efforts in land speculation, see pages 404 and 450–452.)

Leopold, A. 1949. *A Sand County Almanac*. New York: Oxford University Press.

Leopold, E. B. 2016. *Stories from the Leopold Shack: Sand County Revisited*. New York: Oxford University Press.

Meine, C. 1988. *Aldo Leopold: His Life and Work*. Madison, WI: University of Wisconsin Press.

Raghavan, M., et al. 2015. Genomic evidence for the Pleistocene and recent population history of Native Americans. *Science* 349: 791–792.

Strayer, D. L., J. A. Downing, W. R. Haag, T. L. King, J. B. Layzer, T. J. Newton, and S. J. Nichols. 2004. Changing perspectives on pearly mussels, North America's most imperiled animals. *BioScience* 54: 429–439.

On the people of King Island

www.kawerak.org/tribalHomePages/kingIsland/index.html

Bible passages on the Golden Rule: Matthew 7:12 as well as Matthew 22: 39–40, Mark 12:31, and Luke 10: 25–28

TREES

On fossil fuel use, climate change, and the impact of climate change on natural disasters

British Petroleum PLC. 2007. BP Statistical Review of World Energy. www.bp.com/statisticalreview

Emanuel, K. 2005. Increased destructiveness of tropical cyclones over the past 30 years. *Nature* 436: 686–688.

Frame, D. J., and D. A. Stone. 2013. Assessment of the first consensus prediction on climate change. *Nature Climate Change* 3: 357–359.

Hamann, A., and T. Wang. 2006. Potential effects of climate change on ecosystem and tree species distribution in British Columbia. *Ecology* 87: 2773–2786.

IPCC. 2013. Summary for policymakers. In *Climate Change 2013: The Physical Science Basis*, Contribution of Working Group I to the Fifth Assessment Report of the Intergovernmental Panel on Climate Change. Cambridge and New York: Cambridge University Press.

Karl, T. R., et al. 2015. Possible artifacts of data biases in the recent global surface warming hiatus. *Science* 348: 1469–1472.

Kelly, R., M. L. Chipman, P. E. Higuera, I. Stefanova, L. B. Brubaker, and F. S. Hu. 2013. Recent burning of boreal forests exceeds fire regime of the past 10,000 years. *Proceedings of the National Academy of Sciences* 110: 13055–13060.

Mack, M. C., et al. 2011. Carbon loss from an unprecedented Artic tundra wildfire. *Nature* 475: 489–492.

Marotzke, J., and P. M. Forster. 2015. Forcing, feedback and internal variability in global temperature trends. *Nature* 517: 565–570.

Pall, P., et al. 2011. Anthropogenic greenhouse gas contribution to flood risk in England and Wales in autumn 2000. *Nature* 470: 382–385.

Rahmstorf, S., et al. 2007. Recent climate observations compared to projections. *Nature* 316: 709.

Saunders, M. A., and A. S. Lea. 2008. Large contribution of sea surface warming to recent increase in Atlantic hurricane activity. *Nature* 451: 557–560.

Snover, A. K., G. S. Mauger, L. C. Whitely Binder, M. Krosby, and I. Tohver. 2013. *Climate Change Impacts and Adaptation in Washington State: Technical Summaries for Decision Makers*. State of Knowledge

Report prepared for the Washington State Department of Ecology. Seattle: Climate Impacts Group, University of Washington.

Van Vuuren, D. P., et al. 2011. The representative concentration pathways: An overview. *Climatic Change* 109: 5–31.

Webster, P. J., G. J. Holland, J. A. Curry, and H.-R. Chang. 2005. Changes in tropical cyclone number, duration, and intensity in a warming environment. *Science* 309: 1844–1846.

Westerling, A. L., H. G. Hidalgo, D. R. Cayan, and T. W. Swetnam. 2006. Warming and earlier spring increase western U.S. forest wildfire activity. *Science* 313: 940–943.

Young, I. R., S. Zieger, and A. V. Babanin. 2011. Global trends in wind speed and wave height. *Science* 332: 451–455.

On acidification and ozone depletion

Stoddard, J. L., et al. 1999. Regional trends in recovery from acidification in North American and Europe. *Nature* 401: 575–578.

Weatherhead, E. C., and S. B. Andersen. 2006. The search for signs of recovery of the ozone layer. *Nature* 441: 39–45.

For updates on progress in the United States toward lowering gasoline emissions that cause acid rain: www.epa.gov/airmarkt/progress/ARP.

SALMON

On the biology of salmon

Brett, J. R., and C. Groot. 1963. Some aspects of olfactory and visual responses in Pacific salmon. *Journal of the Fisheries Research Board of Canada* 20: 287–303.

Bystriansky, J. S., and P. M. Schulte. 2011. Changes in gill H+-ATPase and Na+/K+-ATPase expression and activity during freshwater acclimation of Atlantic salmon (*Salmo salar*). *Journal of Experimental Biology* 214: 2435–2442.

Cheng, C., and I. N. Flamarique. 2004. New mechanism for modulating colour vision. *Nature* 428: 279.

Craig, J. K., and C. J. Foote. 2001. Countergradient variation and secondary sexual color: Phenotype convergence promotes genetic divergence in carotenoid use between sympatric anadromous and nonanadromous morphs of sockeye salmon (*Oncorhynchus nerka*). *Evolution* 55: 380–391.

Crespi, B. J., and R. Teo. 2002. Comparative phylogenetic analysis of the evolution of semelparity and life history in salmonid fishes. *Evolution* 56: 1008–1020.

Dann, S. G., W. T. Allison, D. B. Levin, J. S. Taylor, and C. W. Hawryshyn. 2004. Salmonid opsin sequences undergo positive selection and indicate an alternate evolutionary relationship in *Oncorhynchus*. *Journal of Molecular Evolution* 58: 400–412.

Drake, D. C., R. J. Naiman, and J. S. Bechtold. 2006. Fate of nitrogen in riparian forest soils and trees: An ^{15}N tracer study simulating salmon decay. *Ecology* 87: 1256–1266.

Flamarique, I. N. 2000. The ontogeny of ultraviolet sensitivity, cone disappearance and regeneration in the sockeye salmon *Oncorhynchus nerka*. *Journal of Experimental Biology* 203: 1161–1172.

Flamarique, I. N., and C. W. Hawryshyn. 1996. Retinal development and visual sensitivity of young Pacific sockeye salmon (*Oncorhynchus nerka*). *Journal of Experimental Biology* 199: 869–882.

Fleming, I. A., and M. R. Gross. 1984. Breeding competition in a Pacific salmon (coho: *Onchrhynchus kisutch*): Measures of natural and sexual selection. *Evolution* 48: 637–657.

Foote, C. J., G. S. Brown, and C. C. Wood. 1997. Spawning success of males using alternative mating tactics in sockeye salmon, *Oncorhynchus nerka*. *Canadian Journal of Fisheries and Aquatic Sciences* 54: 1785–1795.

Foote, C. J., C. C. Wood, and R. E. Withler. 1989. Biochemical genetic comparison of sockeye salmon and kokanee, the anadromous and nonanadromous forms of *Oncorhynchus nerka*. *Canadian Journal of Fisheries and Aquatic Sciences* 467: 149–158.

Garner, S. R., B. D. Neff, and M. A. Bernards. 2010. Dietary carotenoid levels affect carotenoid and retinoid allocation in female Chinook salmon *Oncorhynchus tshawytscha*. *Journal of Fish Biology* 76: 1474–1490.

Gende, S. M., R. T. Edwards, M. F. Willson, and M. S. Wipfli. 2002. Pacific salmon in aquatic and terrestrial ecosystems. *BioScience* 52: 917–928.

Heath, D. D., R. H. Devlin, J. W. Heath, and G. K. Iwama. 1994. Genetic, environmental and interaction effects on the incidence of jacking in *Oncorhynchus tshawytscha* (Chinook salmon). *Heredity* 72: 146–154.

LeMay, M. A., and M. A. Russello. 2015. Genetic evidence for ecological divergence in kokanee salmon. *Molecular Ecology* 24: 798–811.

Lohmann, K. J., N. F. Putnam, and C.M.F. Lohmann. 2008. Geomagnetic imprinting: A unifying hypothesis of long-distance natal homing in salmon and sea turtles. *Proceedings of the National Academy of Sciences* 105: 19096–19101.

Madsen, S. S., P. Kiilerich, and C. K. Tipsmark. 2009. Multiplicity of expression of Na+/K+-ATPase a-subunit isoforms in the gill of Atlantic salmon (*Salmo salar*): Cellular localization and absolute quantification in response to salinity change. *Journal of Experimental Biology* 212: 79–88.

McCormick, S. D., A. M. Regish, and A. K. Christensen. 2009. Distinct freshwater and seawater isoforms of Na+/K+-ATPase in gill chloride cells of Atlantic salmon. *Journal of Experimental Biology* 212: 3994–4001.

Moore, J. W. 2006. Animal ecosystem engineers in streams. *BioScience* 56: 237–246.

Palmer, L. M., M. Deffenbaugh, and A. F. Mensinger. 2005. Sensitivity of the anterior lateral line to natural stimuli in the oyster toadfish. *Journal of Experimental Biology* 208: 3441–3450.

Quinn, T. P., I. J. Stewart, and C. P. Boatright. 2006. Experimental evidence of homing to site of incubation by mature sockeye salmon, *Oncorhynchus nerka. Animal Behaviour* 72: 941–949.

Shrimpton, J. M., et al. 2005. Ionoregulatory changes in different populations of maturing sockeye salmon *Oncorhynchus nerka* during ocean and river migration. *Journal of Experimental Biology* 208: 4069–4078.

Taylor, E. B., C. J. Foote, and C. C. Wood. 1996. Molecular genetic evidence for parallel life-history evolution within a Pacific salmon (sockeye salmon and kokanee, *Oncorhynchus nerka*). *Evolution* 50: 401–416.

Uchida, K., T. Kaneko, K. Yamauchi, and T. Hirano. 1996. Morphometrical analysis of chloride cell activity in the gill filaments and lamellae and changes in Na+, K+-ATPase activity during seawater adaptation in chum salmon fry. *Journal of Experimental Zoology* 276:193–200.

On salmon conservation

Araki, H., B. Cooper, and M. S. Blouin. 2007. Genetic effects of captive breeding cause a rapid, cumulative fitness decline in the wild. *Science* 318: 100–103.

Beamish, R. J., and C. Mahnken. 2001. A critical size and period hypothesis to explain natural regulation of salmon abundance and the linkage to climate and climate change. *Progress in Oceanography* 49: 423–437.

Beaugrand, G., P. C. Reid, F. Ibañez, J. A. Lindley, and M. Edwards. 2002. Reorganization of North Atlantic marine copepod biodiversity and climate. *Science* 296: 1692–1694.

Brinckman, J. 2002. Cost of hatchery salmon careens from $14 to $530 per fish. *Oregonian*, November 12, 2002.

Center for Whale Research. 2015. Southern resident killer whales. www.whaleresearch.com/#!orca-population/cto2

Costello, C., S. D. Gaines, and J. Lynham. 2008. Can catch shares prevent fisheries collapse? *Science* 321: 1678–1681.

Elsner, M. M., et al. 2010. Implications of 21st century climate change for the hydrology of Washington state. *Climatic Change* 102: 225–260.

Grebmeier, J. M., et al. 2006. A major ecosystem shift in the northern Bering Sea. *Science* 311: 1461–1464.

Heath, D. D., J. W. Heath, C. A. Bryden, R. M. Johnson, and C. W. Fox. 2003. Rapid evolution of egg size in captive salmon. *Science* 299: 1738–1740.

Isaak, D. J., S. Wollrab, D. Horan, and G. Chandler. Climate change effects on steam and river temperatures across the northwest U.S. from 1980–2009 and implications for salmonid fishes. *Climatic Change* 113: 499–524.

Kilduff, D. P., E. Di Lorenzo, L. W. Botsford, and S.L.H. Teo. 2015. Changing central Pacific El Niños reduce stability of North American salmon survival rates. *Proceedings of the National Academy of Sciences* 112: 10962–10966.

Mackas, D. L., S. Batten, and M. Trudel. 2007. Effects on zooplankton of a warmer ocean: Recent evidence from the Northeast Pacific. *Progress in Oceanography* 75: 223–252.

Mantua, N., I. Tohver, and A. Hamlet. 2009. Impacts of climate change on key aspects of freshwater salmon habitat in Washington state. Chapter 6 in *The Washington Climate Change Impacts Assessment: Evaluating Washington's Future in a Changing Climate*. Seattle: Climate Impacts Group, University of Washington.

Montgomery, D. 2003. *King of Fish*. Boulder, CO: Westview Press.

Morita, K. 2014. Japanese wild salmon research: Toward a reconciliation between hatchery and wild salmon management. *NPAFC Newsletter* 35: 4–14.

North Atlantic Salmon Conservation Organization. 2008. Implementation plan for Iceland. www.nasco.int/implementation_plans_cycle1.html

Radtke, H. D., and C. N. Carter. 2009. Economic effects and social implications from federal Mitchell act funded hatcheries. Oregon City, OR: Native Fish Society.

Sydeman, W. J., et al. 2014. Climate change and wind intensification in coastal upwelling ecosystems. *Science* 345: 77–80.

Thomas, M. K., C. T. Kremer, C. A. Klausmeier, and E. Litchman. 2012. A global pattern of thermal adaptation in marine phytoplankton. *Science* 338: 1085–1088.

PLANTING SEASON

On deforestation in North America

Williams, M. 1989. *Americans and Their Forests: A Historical Geography.* New York and Cambridge, UK: Cambridge University Press.

On global deforestation

Bond-Lamberty, B., S. D. Peckham, D. E. Ahl, and S. T. Gower. 2007. Fire as the dominant driver of central Canadian boreal forest carbon balance. *Nature* 450: 89–92.

Brando, P. M., et al. 2014. Abrupt increases in Amazonian tree mortality due to drought-fire interactions. *Proceedings of the National Academy of Sciences* 111: 6347–6352.

Chiarello, A. G. 1999. Effects of fragmentation of the Atlantic forest on mammal communities in south-eastern Brazil. *Biological Conservation* 89: 71–82.

Cochrane, M. A. 2003. Fire science for rainforests. *Nature* 421: 913–919.

FAO. 2011. Global forest land-use change from 1990–2005. www.fao.org/forestry/fra/remotesensingsurvey/en/

Hansen, M. C., et al. 2008. Humid tropical forest clearing from 2000 to 2005 quantified by using multitemporal and multiresolution remotely sensed data. *Proceedings of the National Academy of Sciences* 105: 9439–9444.

Hansen, M. C., et al. 2013. High-resolution global maps of 21st-century forest cover change. *Science* 342: 850–853.

Kaplan, J. O., K. M. Krumhardt, and N. Zimmerman. 2009. The prehistoric and preindustrial deforestation of Europe. *Quaternary Science Reviews* 28: 3016–3034.

Kelly, R., M. L. Chipman, P. E. Higuera, I. Stefanova, L. B. Brubaker, and F. S. Hu. 2013. Recent burning of boreal forests exceeds fire regime limits of the past 10,000 years. *Proceedings of the National Academy of Sciences* 110: 13055–13060.

Laporte, N. T., J. A. Stabach, R. Grosch, T. S. Lin, and S. J. Goetz. 2007. Expansion of industrial logging in Central Africa. *Science* 316: 1451.

Laurance, W. F. 1988. A crisis in the making: Responses of Amazonian forests to land use and climate change. *Trends in Ecology and Evolution* 13: 411–415.

Laurance, W. F., et al. 2001. The future of the Brazilian Amazon. *Science* 291: 438–439.

Laurance, W. F., et al. 2002. Ecosystem decay of Amazonian forest fragments: A 22-year investigation. *Conservation Biology* 16: 605–618.

Laurance, W. F., et al. 2006. Rapid decay of tree-community composition in Amazonian forest fragments. *Proceedings of the National Academy of Sciences* 103: 19010–19014.

Laurance, W. F., et al. 2014. Apparent environmental synergism drives the dynamics of Amazonian forest fragments. *Ecology* 95: 3018–3026.

Lewis, S. L., D. P. Edwards, and D. Galbraith. 2015. Increasing human dominance of tropical forests. *Science* 349: 827–832.

Malhi, Y., J. T. Roberts, R. A. Betts, T. J. Killeen, W. Li, and C. A. Nobre. 2008. Climate change, deforestation, and the fate of the Amazon. *Science* 319: 169–172.

Parvianen, J. 2005. Virgin and natural forests in the temperate zone of Europe. *Forest Snow and Landscape Research* 79: 9–18.

Rodrigues, A.S.L., R. M. Eweers, L. Parry, C. Souza Jr., A. Veríssimio, and A. Balmford. 2009. Boom-and-bust development patterns across the Amazon deforestation frontier. *Science* 324: 1435–1437.

Sodhi, N. S., L. P. Koh, B. W. Brook, and P.K.L. Ng. 2004. Southeast Asian biodiversity: An impending disaster. *Trends in Ecology and Evolution* 19: 654–660.

Tropek, R., et al. 2013. Comment on "High-resolution global maps of 21st-century forest cover change." *Science* 344: 981.

Westerling, A. L., H. G. Hidalgo, D. R. Cayan, and T. W. Swetnam. 2006. Warming and earlier spring increase western U.S. forest wildfire activity. *Science* 313: 940–943.

Wilcove, D. S., X. Giam, D. P. Edwards, B. Risher, and L. P. Koh. 2014. Novjot's nightmare revisited: Logging, agriculture, and biodiversity in Southeast Asia. *Trends in Ecology and Evolution* 28: 531–540.

Bible passage referring to Solomon's Temple and the cedars of Lebanon: I Kings 5.

On the impacts of climate change on forests

Fang, J., et al. 2014. Evidence for environmentally enhanced forest growth. *Proceedings of the National Academy of Sciences* 111: 9527–9532.

Isbell, F., P. B. Reich, D. Tilman, S. E. Hobbie, S. Polasky, and S. Binder. 2013. Nutrient enrichment, biodiversity loss, and consequent declines in ecosystem productivity. *Proceedings of the National Academy of Sciences* 110: 11911–11916.

McMahon, S. M., G. G. Parker, and D. R. Miller. 2010. Evidence for a recent increase in forest growth. *Proceedings of the National Academy of Sciences* 107: 3611–3615.

On restoration efforts

Boydak, M. 1996. *Ecology and Silviculture of Cedar of Lebanon* (Cedrus libani *A. Rich.*) *and Conservation of Its Natural Forests*. Istanbul: Orman Bakanligi Yayin.

Boydak, M. 2003. Regeneration of Lebanon cedar (*Cedrus libani* A. Rich.) on karstic lands in Turkey. *Forest Ecology and Management* 178: 231–243.

Cetin, H., Y. Kurt, K. Isik, and A. Yanikoglu. 2009. Larvicidal effect of *Cedrus libani* seed oils on mosquito *Culex pipiens*. *Pharmaceutical Biology* 47: 665–668.

Hajar, L., L. François, C. Khater, I. Jomaa, M. Déqué, and R. Cheddadi. 2010. *Cedrus libani* (A. Rich.) distribution in Lebanon: Past, present, and future. *Comptes Rendus Biologies* 333: 622–630.

Leopold, A. C. 2005. Toward restoration of a wet tropical forest in Costa Rica: A ten-year report. *Ecological Restoration* 23: 230–233.

Leopold, A. C., and J. Salazar. 2008. Understory species richness during restoration of wet tropical forest in Costa Rica. *Ecological Restoration* 26: 22–24.

Little, J. B. 2008. Regrowing Borneo's rainforest—Tree by tree. *Scientific American Earth 3.0* 18: 64–71.

Xu, J. 2011. China's new forests aren't as green as they seem. *Nature* 477: 371.

Zhang, P., et al. 2000. China's forest policy for the 21st century. *Science* 288: 2135–2136.

For updates on the reforestation projects referred to in this chapter
www.plant-for-the-planet-billiontreecampaign.org/
www.tasikoki.org/

On reforestation in the Pacific Northwest, especially tree-fungal associations

Barto, E. K., J. D. Weidenhamer, D. Cipollini, and M. C. Rillig. 2012. Fungal superhighways: Do common mycorrhizal networks enhance below ground communication? *Trends in Plant Science* 17: 633–637.

Behie, S. W., and M. J. Bidochka. 2014. Nutrient transfer in plant-fungal symbioses. *Trends in Ecology and Evolution* 19: 734–740.

Brachmann, A., and M. Parniske. 2006. The most widespread symbiosis on Earth. *PLoS Biology* 4: e239–e240.

Brundrett, M. 2004. Diversity and classification of mycorrhizal associations. *Biological Reviews* 79: 473–495.

Bücking, H., and W. Heyser. 2001. Microautoradiographic localization of phosphate and carbohydrates in mycorrhizal roots of *Populus tremula* × *Populus alba* and the implications for transfer processes in ectomycorrhizal associations. *Tree Physiology* 21: 101–107.

Maltz, M. R., and K. K. Treseder. 2015. Sources of inocula influence mycorrhizal colonization of plants in restoration projects: A meta-analysis. *Restoration Ecology* 23: 625–634.

Miller, S. L., C. D. Koo, and R. Molina. 1992. Early colonization of red alder and Douglas fir by ectomycorrhizal fungi and *Frankia* in soils from the Oregon coast range. *Mycorrhiza* 2: 53–61.

Pfeffer, P. E., D. D. Douds Jr., H. Bücking, D. P. Schwartz, and Y. Shachar-Hill. 2004. The fungus does not transfer carbon to or between roots in an arbuscular mycorrhizal symbiosis. *New Phytologist* 163: 617–627.

Riege, D. A., and R. del Moral. Differential tree colonization of old fields in a temperate rain forest. *American Midland Naturalist* 151: 251–264.

Selosse, M.-A., F. Richard, X. He, and S. W. Simard. 2006. Mycorrhizal networks: Des liaisons dangereuses? *Trends in Ecology and Evolution* 21: 621–628.

Simard, S. W., M. D. Jones, D. M. Durall, D. A. Perry, D. D. Myrold, and R. Molina. 1997. Reciprocal transfer of carbon isotopes between ectomycorrhizal *Betula papyrifera* and *Pseudotsuga menziesii*. *New Phytologist* 137: 529–542.

Simard, S. W., D. A. Perry, M. D. Jones, D. D. Myrold, D. M. Durall, and R. Molina. 1997. Net transfer of carbon between ectomycorrhizal tree species in the field. *Nature* 388: 579–582.

BLOOD, SWEAT, TEARS

Bais, H. P., R. Vepachedu, S. Gilroy, R. J. Callaway, and J. M. Vivanco. 2003. Allelopathy and exotic plant invasion: From molecules and genes to species interactions. *Science* 301: 1377–1380.

Blair, A. C., S. J. Nissen, G. R. Brunk, and R. A. Hufbauer. 2006. A lack of evidence for an ecological role of the putative allelochemical

(±)-catechin in spotted knapweed invasion success. *Journal of Chemical Ecology* 32: 2327–2331.

Burke, D. J. 2008. Effects of *Alliaria petiolata* (garlic mustard; Brassicaceae) on mycorrhizal colonization and community structure in three herbaceous plants in a mixed deciduous forest. *American Journal of Botany* 95: 1416–1425.

Chisholm, S. T., G. Coaker, B. Day, and B. J. Staskawicz. 2006. Host-microbe interactions: Shaping the evolution of the plant immune response. *Cell* 124: 803–814.

Dangl, J. L., and J.D.G. Jones. 2001. Plant pathogens and integrated defense responses to infection. *Nature* 411: 826–833.

DeWalt, S. J., J. S. Denslow, and K. Ickes. 2004. Natural-enemy release facilitates habitat expansion of the invasive tropical shrub *Clidemia hirta*. *Ecology* 85: 471–483.

Duncan, R. P., A. G. Boyer, and T. M. Blackburn. 2013. Magnitude and variation of prehistoric bird extinctions in the Pacific. *Proceedings of the National Academy of Sciences* 110: 6436–6441.

Frick, W. F., et al. 2010. An emerging disease causes regional population collapse of a common North American bat species. *Science* 329: 679–682.

Griffin, G. 2000. Blight control and restoration of the American chestnut. *Journal of Forestry* 98: 22–27.

Heil, M., and J.C.S. Bueno. 2007. Within-plant volatiles leads to induction and priming of an indirect plant defense in nature. *Proceedings of the National Academy of Sciences* 104: 5467–5472.

Heil, M., and J. Ton. 2008. Long-distance signaling in plant defense. *Trends in Plant Science* 13: 264–272.

Hepting, G. H. 1974. Death of the American chestnut. *Journal of Forest History* 18: 60–67.

Jones, J.D.G., and J. L. Dangl. 2006. The plant immune system. *Nature* 444: 322–329.

LaDeau, S. L., A. M. Kilpatrick, and P. P. Marra. 2007. West Nile virus emergence and large-scale declines of North American bird populations. *Nature* 447: 710–713.

Martel, A., et al. 2014. Recent introduction of a chytrid fungus endangers Western Palearctic salamanders. *Science* 346: 630–631.

McCormick, A. C., S. B. Unsicker, and J. Gershenzon. 2012. The specificity of herbivore-induced plant volatiles in attracting herbivore enemies. *Trends in Plant Science* 17: 303–310.

Mitchell, C. E., and A. G. Power. 2003. Release of invasive plants from fungal and viral pathogens. *Nature* 421: 625–627.

Pimentel, D., R. Zuniga, and D. Morrison. 2004. Update on the environmental and economic costs associated with alien-invasive species in the United States. *Ecological Economics* 52: 273–288.

Steadman, D. W. 1995. Prehistoric extinctions of Pacific Island birds: Biodiversity meets zooarchaeology. *Science* 267: 1123–1131.

Stinson, K. A., et al. 2006. Invasive plant suppresses the growth of native tree seedlings by disrupting belowground mutualisms. *PLoS Biology* 4: e140.

Torchin, M. E., K. D. Laffery, A. P. Dobson, V. J. McKenzie, and A. M. Kuris. 2003. Introduced species and their missing parasites. *Nature* 421: 628–630.

Van Riper III, C., S. G. van Riper, M. L. Goff, and M. Laird. 1986. The epizootiology and ecological significance of malaria in Hawaiian land birds. *Ecological Monographs* 56: 327–344.

Venter, O., N. N. Brodeur, L. Nemiroff, B. Belland, I. J. Dolinsek, and J. W. A. Grant. 2006. Threats to endangered species in Canada. *BioScience* 56: 903–910.

Vredenburg, V. T., R. A. Knapp, T. S. Tunstall, and C. J. Briggs. 2010. Dynamics of an emerging disease drive large-scale amphibian population extinctions. *Proceedings of the National Academy of Sciences* 107: 9689–9694.

A WORKING FOREST

On human population growth

Gerland, P., et al. 2014. World population stabilization unlikely this century. *Science* 346: 234–237.

United Nations Population Information Network. 2012. *World Population Prospects: 2012 Revision.* New York: United Nations.

For updates on human population size, see the U.S. and World Population Clock published by the U.S. Census Bureau at www.census.gov/popclock/.

On forests as carbon sinks

Cohen, W. B., M. E. Harmon, D. O. Wallin, and M. Fiorella. 1996. Two decades of carbon flux from forests of the Pacific Northwest. *BioScience* 46: 836–844.

Luyssaert, S., et al. 2008. Old-growth forests as global carbon sinks. *Nature* 455: 213–215.

Stephenson, N. L., et al. 2014. Rate of tree carbon accumulation increases continuously with tree size. *Nature* 507: 90–93.

Zhou, G., et al. 2006. Old-growth forests can accumulate carbon in soils. *Science* 314: 1417.

On the "war in the woods" in the Pacific Northwest

Dietrich, W. 1992. *The Final Forest: The Battle for the Last Great Trees of the Pacific Northwest.* Seattle: University of Washington Press. See also the 2010 updated edition.

DAMNATION

On beavers

Castro, J., M. Pollock, C. Jordan, G. Lewallen, and K. Woodruff (eds.). 2015. *The Beaver Restoration Guidebook.* Portland, OR: U.S. Fish and Wildlife Service.

Feinstein, K. 2006. A brief history of the beaver trade. cwh.ucsc.edu/fein-stein/A%20brief%20history%20of%20the%20beaver%20trade.html

Ji, Q., Z.-X. Luo, C.-X. Yuan, and A. R. Tabrum. 2006. A swimming mammaliaform from the Middle Jurassic and ecomorphological diversification of early mammals. *Science* 311: 1123–1127.

Naiman, R. J., C. A. Johnston, and J. C. Kelly. 1988. Alterations of North American streams by beaver. *BioScience* 38: 753–762.

Seton, E. T. 1929. *Lives of Game Animals*, Vol. 4, Part 2, Rodents, Etc. Garden City, NY: Doubleday.

On mass extinctions

Adair, R. K. 2010. Wildfires and animal extinctions at the Cretaceous/Tertiary boundary. *American Journal of Physics* 78: 567–573.

Alvarez, W., F. Asaro, and A. Montanari. 1990. Iridium profile for 10 million years across the Cretaceous-Tertiary boundary at Gubbio (Italy). *Science* 250: 1700–1702.

Benton, M. J. 1995. Diversification and extinction in the history of life. *Science* 268: 52–58.

Clarke, J. A., C. P. Tambussi, J. I. Oriega, G. M. Erickson, and R. A. Ketcham. 2005. Definitive fossil evidence for the extant avian radiation in the Cretaceous. *Nature* 433: 305–308.

Florentin, J.-M., R. Maurrasse, and G. Sen. 1991. Impacts, tsunamis, and the Haitian Cretaceous-Tertiary boundary layer. *Science* 252: 1690–1693.

Heymann, D., L.P.F. Chibante, R. R. Brooks, W. S. Wolbach, and R. E. Smalley. 1994. Fullerenes in the Cretaceous-Tertiary boundary layer. *Science* 265: 645–647.

Jablonski, D., and D. M. Raup. 1995. Selectivity of end-Cretaceous marine bivalve extinctions. *Science* 268: 389–391.

Quental, T. B., and C. R. Marshall. 2013. How the Red Queen drives terrestrial mammals to extinction. *Science* 341: 290–292.

Raup, D. 1986. Biological extinction in Earth history. *Science* 231: 1528–1533.

Schulte, P., et al. 2010. The Chicxulub asteroid impact and mass extinction at the Cretaceous-Paleogene boundary. *Science* 327: 1214–1218.

Schultz, P. H., and S. D'Hondt. 1996. Cretaceous-Tertiary (Chicxulub) impact angle and its consequences. *Geology* 24: 963–967.

Swisher, C. C. III, et al. 1992. Coeval ^{40}Ar/^{39}Ar ages of 65.0 million years ago from Chicxulub crater melt rock and Cretaceous-Tertiary boundary tektites. *Science* 257: 954–958.

Turco, R. P., O. B. Toon, T. P. Ackerman, J. B. Pollack, and C. Sagan. 1990. Climate and smoke: An appraisal of nuclear winter. *Science* 247: 166–176.

On human evolution

Antón, S. C., R. Potts, and L. C. Aiello. 2014. Evolution of early Homo: An integrated biological perspective. *Science* 345: 45.

Benazzi, S., et al. 2015. The makers of the Protoaurignacian and implications for Neandertal extinction. *Science* 348: 793–796.

Bowler, J. M. 2003. New ages for human occupation and climatic change at Lake Mungo, Australia. *Nature* 421: 837–840.

Brown, P., T. Sutikna, M. J. Morwood, R. P. Soejono, Jatmiko, E. W. Saptomo, and R. A Due. 2003. A new small-bodied hominin from the late Pleistocene of Flores, Indonesia. *Nature* 431: 1055–1061.

Francalacci, P., et al. 2013. Low-pass DNA sequencing of 1200 Sardinians reconstructs European Y-chromosome phylogeny. *Science* 341: 565–569.

Goebel, T., M. R. Waters, and D. H. O'Rourke. 2008. The Late Pleistocene dispersal of modern humans in the Americas. *Science* 319: 1497–1502.

Higham, T., et al. 2014. The timing and spatiotemporal patterning of Neanderthal disappearance. *Nature* 512: 306–309.

Joordens, J.C.A., et al. 2015. *Homo erectus* at Trinil on Java used shells for tool production and engraving. *Nature* 518: 228–231.

McDougall, I., F. H. Brown, and J. G. Reagle. 2005. Stratigraphic placement and age of modern humans from Kibish, Ethiopia. *Nature* 433: 733–736.

Mellars, P. 2006. A new radiocarbon revolution and the dispersal of modern humans in Eurasia. *Nature* 439: 931–935.

Poznik, G. P., et al. 2013. Sequencing Y chromosomes resolves discrepancy in time to common ancestor in males versus females. *Science* 341: 562–565.

Summerhayes, G. R., et al. 2010. Human adaptation and plant use in highland New Guinea 49,000 to 44,000 years ago. *Science* 330: 78–81.

White, T. D., et al. 2003. Pleistocene *Homo sapiens* from Middle Awash, Ethiopia. *Nature* 423: 742–747.

Zhu, R. X., et al. 2001. Earliest presence of humans in northeast Asia. *Nature* 413: 413–417.

On the current mass extinction

Barnosky, A. D., et al. 2011. Has the Earth's sixth mass extinction already arrived? *Nature* 471: 51–57.

Costello, M. J., R. M. May, and N. E. Stork. 2013. Can we name Earth's species before they go extinct? *Science* 339: 413–416.

Duncan, R. P., A. G. Boyer, and T. M. Blackburn. 2013. Magnitude and variation of prehistoric bird extinctions in the Pacific. *Proceedings of the National Academy of Sciences* 110: 6436–6441.

Lorenzen, E. D., et al. 2011. Species-specific responses of Late Quaternary megafauna to climate and humans. *Nature* 479: 359–364.

Prescott, G. W., D. R. Williams, A. Balmford, R. E. Green, and A. Manica. 2012. Quantitative analysis of the role of climate and people in explaining late Quarternary megafaunal extinctions. *Proceedings of the National Academy of Sciences* 109: 4527–4531.

Steadman, D. W. 1995. Prehistoric extinctions of Pacific Island birds: Biodiversity meets zooarchaeology. *Science* 267: 1123–1131.

WILD THINGS

Bratman, G. N., J. P. Hamilton, K. S. Hahn, G. C. Daily, and J. J. Gross. 2015. Nature experience reduces rumination and subgenual prefrontal cortex activation. *Proceedings of the National Academy of Sciences* 112: 8567–8572.

Cavalli-Sforza, L. L., and M. W. Feldman. 2003. The application of molecular genetic approaches to the study of human evolution. *Nature* 33: 266–275.

Dadvand, P., et al. 2015. Green spaces and cognitive development in primary schoolchildren. *Proceedings of the National Academy of Sciences* 112: 7937–7942.

Gaitán, J. J., et al. 2015. Plant species richness and shrub cover attenuate drought effects on ecosystem functioning across Patagonian rangelands. *Biology Letters* 10: 20140673.

Hautier, Y., D. Tilman, F. Isbell, E. W. Seabloom, E. T. Borer, and P. B. Reich. 2015. Anthropogenic environmental changes affect ecosystem stability via biodiversity. *Science* 348: 336–340.

Li, J. Z., et al. 2008. Worldwide human relationships inferred from genome-wide patterns of variation. *Science* 319: 1100–1104.

Outdoor Foundation. 2011. *Outdoor Recreation Participation Report.* Boulder, CO: Outdoor Foundation.

Paabo, S. 2014. The human condition—A molecular approach. *Cell* 157: 216–226.

Pergrams, O.R.W., and P. A. Zaradic. 2008. Evidence for a fundamental and pervasive shift away from nature-based recreation. *Proceedings of the National Academy of Sciences* 105: 2295–2230.

Reich, P. B., et al. 2012. Impacts of biodiversity loss escalate through time as redundancy fades. *Science* 336: 589–592.

St. Leger, L. 2003. Health and nature—New challenges for health promotion. *Health Promotion International* 18: 173–175.

Tilman, D., and J. A. Downing. 1994. Biodiversity and stability in grasslands. *Nature* 367: 363–365.

Tilman, D., P. B. Reich, and F. Isbell. 2012. Biodiversity impacts ecosystem productivity as much as resources, disturbance, or herbivory. *Proceedings of the National Academy of Sciences* 109: 10394–10397.

Ulrich, R. S. 1984. View through a window may influence recovery from surgery. *Science* 224: 420–422.

U.S. Fish and Wildlife Service. 2011. *National Survey of Fishing, Hunting, and Wildlife-Associated Recreation* (2014 update). Washington, DC: U.S. Department of the Interior.

Vander Heijden, M.G.A., et al. 1998. Mycorrhizal fungal diversity determines plant biodiversity, ecosystem variability and productivity. *Nature* 396: 69–72.

A NATURAL LIFE

Ackman, D. 2001. How big is porn? *Forbes* blog, May 25.

Adams, J. T. 1931. *The Epic of America*. New York: Atlantic Monthly Press.

Baumeister, R. F., K. D. Vohs, J. L. Aaker, and E. N. Garbinsky. 2013. Some key differences between a happy life and a meaningful life. *Journal of Positive Psychology* 8: 505–516.

Bradley, N. L., A. C. Leopold, J. Ross, and W. Huffaker. 1999. Phenological changes reflect climate change in Wisconsin. *Proceedings of the National Academy of Sciences* 96: 9701–9704.

Brooks, A. C. 2013. A formula for happiness. *New York Times*, December 15, p. SR1.

Brooks, A. C. 2014. Love people, not pleasure. *New York Times Sunday Review*, July 20, p. SR1.

Brooks, D. 2015. When cultures shift. *New York Times*, April 17, p. A31.

Dallman, M. F., N. C. Pecoraro, and S. E. la Fleur. 2005. Chronic stress and comfort foods: Self-medication and abdominal obesity. *Brain, Behavior, and Immunity* 19: 275–280.

Frey, B. S., and A. Stutzer. 2001. What can economists learn from happiness research? CESifo Working Paper No. 503.

Kahneman, D., and A. Deaton. 2010. High income improves evaluation of life but not emotional well-being. *Proceedings of the National Academy of Sciences* 107: 16489–16493.

Lerner, J., J. Mackey, and F. Casey. 2007. What's in Noah's wallet? Land conservation spending in the United States. *BioScience* 57: 419–423.

Markou, A., T. R. Kosten, and G. F. Koob. 1998. Neurobiological similarities in depression and drug dependence: A self-medication hypothesis. *Neuropsychopharmacology* 18: 135–174.

Meule, A. 2011. How prevalent is "food addiction?" *Frontiers in Psychiatry* 2: 1–4.

NCADD. 2014. www.ncadd.org/about-addiction/faq/facts-about-drugs

Ogden, C. L., M. D. Carroll, B. K. Kit, and K. M. Flagal. 2014. Prevalence of childhood and adult obesity in the United States, 2011–2012. *Journal of the American Medical Association* 311: 806–814.

Pratt, L. S., D. J. Brody, and Q. Gu. 2011. Antidepressant use in persons aged 12 and over: United States, 2005–2008. NCHS Data Brief No. 76.

Weissmann, J. 2011. The Halloween economy: $2 billion in candy, $300 million in pet costumes. *The Atlantic*, October 28.

Wolfe, Tom. 1976. The "Me" decade and the Third Great Awakening. *New York* magazine, August 23.

Bible passages referring to the difficulty of a rich person entering the Kingdom of God: Matthew 19: 23–24.

Index